Technology and the Growth of Civilization

More information about this series at http://www.springer.com/series/4097

Giancarlo Genta • Paolo Riberi

Technology and the Growth of Civilization

Published in association with
Praxis Publishing
Chichester, UK

Giancarlo Genta
Department of Mechanics
Politecnico di Torino
Torino, Italy

Paolo Riberi
Centallo, Italy

SPRINGER-PRAXIS BOOKS IN POPULAR SCIENCE

ISSN 2626-6113   ISSN 2626-6121   (electronic)
Popular Science
ISBN 978-3-030-25582-4   ISBN 978-3-030-25583-1   (eBook)
Springer Praxis Books
https://doi.org/10.1007/978-3-030-25583-1

© Springer Nature Switzerland AG 2019
This work is subject to copyright. All rights are reserved by the Publisher, whether the whole or part of the material is concerned, specifically the rights of translation, reprinting, reuse of illustrations, recitation, broadcasting, reproduction on microfilms or in any other physical way, and transmission or information storage and retrieval, electronic adaptation, computer software, or by similar or dissimilar methodology now known or hereafter developed.
The use of general descriptive names, registered names, trademarks, service marks, etc. in this publication does not imply, even in the absence of a specific statement, that such names are exempt from the relevant protective laws and regulations and therefore free for general use.
The publisher, the authors, and the editors are safe to assume that the advice and information in this book are believed to be true and accurate at the date of publication. Neither the publisher nor the authors or the editors give a warranty, express or implied, with respect to the material contained herein or for any errors or omissions that may have been made. The publisher remains neutral with regard to jurisdictional claims in published maps and institutional affiliations.

Cover design: Jim Wilkie
Project Editor: Michael D. Shayler

Cover images used under license from Shutterstock.com

This Springer imprint is published by the registered company Springer Nature Switzerland AG
The registered company address is: Gewerbestrasse 11, 6330 Cham, Switzerland

Contents

Preface by the Authors, with Acknowledgements viii

Preface by Giuseppe Tanzella Nitti xv

Authors' Introduction xx

1 Technology in prehistory 1
   1.1 A technological animal 1
   1.2 Stone implements 5
   1.3 The cognitive revolution 7
   1.4 Fire control 10
   1.5 The Mesolithic Crisis 12
   1.6 The Neolithic Revolution 14
   1.7 The Age of Metals 17

2 From prehistory to history 19
   2.1 Transmission of information 19
   2.2 Technology at the beginning of history 21
   2.3 Simple machines 23
   2.4 Craftsmen and slaves 29
   2.5 The invention of money 35

3 Greek rationality 39
   3.1 A unique phenomenon 39
   3.2 Humans, Gods and technology 41
   3.3 Applied rationality: The origins of medicine 44
   3.4 Greek natural philosophy 46
   3.5 Hellenistic scientific technology 49
   3.6 The end of Hellenistic science 51
   3.7 A Hellenistic industrial revolution? 54

Contents

4 From Abraham to Jesus: The Judeo-Christian rational horizon 57
- 4.1 In the beginning was the *Logos*... 57
- 4.2 The rational God of the Old Testament 59
- 4.3 Judaism and centrifugal thrusts 62
- 4.4 Circular time, the myth of the ages and progress 66
- 4.5 The Christian DNA of the technological West 70
- 4.6 The two souls of Christianity 73

5 The Roman world and the "broken history" 77
- 5.1 The 'Pillars of Heracles' of ancient technology 77
- 5.2 Matter, slaves and machines: A blind alley 79
- 5.3 The structural limitation of the Roman economy 82
- 5.4 Not just shadows: Imperial technology 83
- 5.5 Metallurgy 85
- 5.6 Military technology 86
- 5.7 Construction industry 87
- 5.8 Energy production 90
- 5.9 Agricultural technologies 93
- 5.10 Means of transportation 94

6 The Middle Ages: "Dark ages" or the dawn of technology? 97
- 6.1 The invention of the Dark Ages 97
- 6.2 Four common myths about the early Middle Ages (476–1000) 98
- 6.3 The Legacy of Hellenistic scientific technology 106
- 6.4 Christianity and the rise of the *artes mechanicae* 110
- 6.5 Work and progress, the turning point of monasticism 113
- 6.6 The medieval technological revolution 116
- 6.7 Finance and accounting in the Middle Ages: The dawn of the modern economy 118

7 The beginning of scientific technology 125
- 7.1 Technology at the transition from the Middle Ages to the Modern Age 125
- 7.2 The "theatres of machines" of the Renaissance 128
- 7.3 Overturning the stereotype of the 'vile mechanicist' 130
- 7.4 The birth of modern science 132
- 7.5 The steam engine 134
- 7.6 The dream of flying 137
- 7.7 Electromagnetism 143
- 7.8 Nuclear energy 145
- 7.9 The École Polytechnique 149

8 Technology, capitalism and imperialism: The rise of the West 152
- 8.1 Europe conquering the world 152
- 8.2 Science and empire 157
- 8.3 Progress generates progress: A system based on trust 158
- 8.4 Capitalist theory and the role of technology 161

	8.5	The dark side of the capitalist empire.	164
	8.6	The utilitarian drift of the West.	168
9	**The dark side of technology**		**172**
	9.1	Weapons and dual technologies	172
	9.2	Mass destruction in ancient times.	173
	9.3	Ancient bacteriological wars	175
	9.4	Nuclear weapons.	176
	9.5	Dual technologies	180
10	**Industrial revolutions**		**183**
	10.1	The agricultural revolution	183
	10.2	The First Industrial Revolution.	185
	10.3	The Second Industrial Revolution	192
	10.4	The Third Industrial Revolution	194
	10.5	Industrial Revolution 4.0	199
	10.6	Some considerations on industrial revolutions	203
11	**The irrationalistic constant**		**205**
	11.1	At the origins of magical thought.	205
	11.2	Irrationalism and religion	209
	11.3	The return of Ouroboros.	210
	11.4	The shadows of the Enlightenment.	213
	11.5	Philosophy vs. Technology: The decline of the West?	217
12	**Beyond the horizon**		**223**
	12.1	Future perspectives.	223
	12.2	Beyond the technology (and the economy) of oil	224
	12.3	Electronics and telecommunications	228
	12.4	Beyond Earth	232
	12.5	The second Neolithic Revolution	235
	12.6	Many human species.	236

Epilogue ... 239

Appendix: Chronological tables ... 251

Bibliography ... 255

Index ... 260

Preface by the Authors, with Acknowledgements

This first part of this preface has been written by only one of the authors (the first one, in strict alphabetical order), for the simple reason that it deals with things happened at a time when the second author was still – as they say – *in the mind of God* and, indeed, his parents were still children.

This book has had a gestation of almost half a century and, unlike certain monsters from science fiction B-movies, which remain hidden for geological eras to pop out unexpectedly and cause great havoc, it has been stirring and kicking in my mind all this time.

Looking back, the first, very brief draft was even earlier. I was in high school, and I had just finished reading a science fiction novel, *Iperbole Infinita*, written by Julian Berry – the pen name of an author who would later become one of the greatest screenwriters of Italian cinema, Ernesto Gastaldi. I thank him, in spite of the bad mark in Italian I received on that occasion, for having enthused me about the subject of technological progress. The title of the book alludes transparently to the fact that the trajectory of humankind may not be a parabola (which first ascends and then inexorably descends) but a hyperbola, which continues to rise to infinity.

One teacher of Italian literature asked us to write an essay on a subject that relates vaguely to what is dealt with in this book. I wrote a dozen pages filled to the brim with rhetoric and concluded by quoting, probably incorrectly, the novel by the would-be Berry (the professor had certainly not read it). The result was the only 4/10 of my school career (practically the worst mark you could get for an essay), with a comment which defined what I had written as a mixture of rhetorical absurdities; no Italian correction but a blue line on the final quotation.

It was a much deserved bad mark, for which thanks must go to that professor, (even if I do not remember which of the many teachers of Italian I had in high school he was), because with it he taught me things like avoiding rhetoric and verbosity, to be well researched before writing and to be respectful of my readers by being succinct and to the point.

The idea of writing a book on the importance of technology and engineering crystallized in the 1970s, following many talks with my former Aeronautical Constructions professor, Piero Morelli, who was then becoming my senior colleague (more thanks). He claimed that an engineer is a *'transformer of matter'* and that his role consists of transforming the world to adapt it to humankind.

But the 1970s were a bad time for a book like this: Western civilization was renouncing so many of the achievements of the Third Industrial Revolution, from nuclear energy, to the exploration of space, to supersonic airliners, thus jeopardizing its very identity and its historical function.

Time passed and not even one page had been written. There was only a folder on my computer, which was filled with articles on all the possible topics related to the main theme (that bad mark still burned). In particular, I collected articles that challenged the politically correct views on topics such as global warming, numerical methods and the role of computers, anti-technological issues and the diffusion of pseudo-sciences.

The second author of this book was born at the end of the 1980s, though of course I had no knowledge of this.

In the early years of the new millennium, I talked about this editorial project to my friend Ezio Quarantelli of Lindau publishing house (verbally) and to Clive Horwood of Praxis (by e-mail), which had recently become a part of Springer. Both had published some of my essays, respectively, in Italian and English. They showed interest in publishing the text, even if there was not a single written page or even a proposal.

In the meantime, I read two essays that strongly influenced my way of thinking, confirming and substantiating some ideas, which I had reached somewhat arbitrarily, and modifying others: *Guns, Germs and Steel: The Fates of Human Societies* by Jared Diamond and *Intellectual Impostures* by Alan Sokal and Jean Bricmont. To all of them go further thanks, even if I have never met them in person, except for the last one with whom I exchanged a few words at the end of a lecture in Turin.

Both Giuseppe Tanzella Nitti, an astronomer, priest and theologian, professor of fundamental theology at the Pontificia Università della Santa Croce in Rome and Adjunct Scholar at the Vatican Observatory, and Gregory Benford, a professor of physics at the University of California, Irvine, and famous science fiction author, provided much greater contributions to this book. The former advised me to read the book *The religion of technology: the divinity of man and the spirit of invention* by D. F. Noble, while the latter, enthusiastic about this editorial project, invented the title (for the Italian edition) with his usual imaginative approach.

I devoured Noble's book, which certainly overturned my very conventional approach to the problem of rationality and irrationalism (rationality and science are the fruits of the Renaissance and the Enlightenment, while the irrational and anti-technological approach can be traced back to the Church of the Dark Ages). However, while I found Noble's analysis so convincing that it completely changed my mind, I was in complete disagreement with his conclusions. What for me was, at that point, the great merit of the Christian world, or rather Judeo-Christian, or better still, Greek-Judeo-Christian (as I understood later), for him was the original sin of modern civilization.

Finally, another turning point came when I came in contact with works by Rodney Stark, above all, *How the West Won: The Neglected Story of the Triumph of Modernity*, through my wife Franca (thanks to her will come later, because they are not limited to this) who translated them into Italian for Lindau. Noble's approach was there, although turned upside-down. It was a revelation: I realized that here was what I had always thought, a sort of intuition with which I agreed instinctively, but now it was being discussed by a true historian and demonstrated with all the necessary details. Apart from that, reading his books was like a breath of fresh air, in a world suffocated by politically correct dogmas.

In the meantime, pure chance had entered the long gestation of this book. In 2010 in Jerusalem, where the modern Judeo-Christian civilization had begun, by chance I met a student of ancient literature, with a deep interest in the Gnostic sects of the first centuries of our era.

We became friends, or better, our families became friends, and when he obtained his master's degree in Ancient Philology and Literature, began his degree course in Economics and published his first essays with Lindau publishing house, I proposed him as a participant in writing *Technology and the Growth of Human Civilization*. His cultural background in history – in particular in the history of religions – and then also in economics, would complement my scientific-technological cultural background.

At that point we began to write, and we realized that, after so many years of preparation, the ideas were now clear and the text was almost writing itself.

Apart from the people already mentioned, we should thank a large number of others. Among them:

The late Giovanni (Nanni) Bignami, professor of astrophysics and, among other things, president of the Italian Space Agency, in particular for two lessons he left us. The first is not to let oneself be dominated by one's political ideas – and militancy – when dealing with technological issues which can influence the future of us all. When he was a member of the Governing Board of the Democratic Party, he wrote, together with other famous scientists, a letter to the then secretary of his party in support of the construction of nuclear power stations in Italy, a letter which obviously was immediately trashed. The second is that of being able to look to a distant future, without being too influenced by contingent technical difficulties. We wrote a scientific paper together (for the *Acta Astronautica Journal*) on the nuclear rocket designed by the Nobel laureate Carlo Rubbia, a device which could turn out to be unfeasible for a long time to come, as Les Shephard told me many years ago. Shepard was one of the founders of the International Academy of Astronautics and successor to Arthur C. Clarke at the helm of the British Interplanetary Society, to whom we owe more thanks for his warning to oppose the fashions of the moment and to remain faithful to our ideas in the technical-scientific field. While writing that paper, whenever the technical difficulties were mentioned, Nanni used to say provocatively (in French) *'ce n'est que du beton'* – it's just concrete. He was well aware that, as the great architects of modernism had taught us, starting with Pier Luigi Nervi, reinforced concrete was a science and an art (and, as the recent disaster of the Morandi Bridge in Genoa has once more tragically taught us, the devil hides in the technical details), but we must always look far, in the certainty that the technical difficulties will be resolved sooner or later.

Thanks must also go to colleagues from the Polytechnic University of Turin, and in particular Riccardo Adami, who, during long and very interesting discussions in a restaurant in Kuala Lumpur, introduced me to the works on Hellenistic science and technology by Lucio Russo, which made the contribution of the Greek heritage to the Judeo-Christian civilization much clearer.

Tuvia Fogel, whom we could define as a literary agent and writer, if this definition were not too reductive, showed us in endless discussions (with him it would be better to call them fascinating monologues) that repository of rationality – and also of irrationalism – which is the Jewish component of our culture.

Lorenzo Morello, former FIAT manager and professor of Automotive Engineering, co-author of many books on motor vehicle technology and history, who made us aware of the irrational component in the invention of a technology such as automotive technology, which had little practical use at the beginning, at least in Europe, but merely fulfilled the dreams and desires of a number of eccentrics who wanted a new toy to satisfy their desires for exhibitionism and thrill seeking. Only later did it become a means of transportation which has proven to be irreplaceable.

Lino Sacchi, former professor of geology at the University of Turin, who introduced us to the book by Huntington, which was essential to clarify some aspects related to what we mean by modernization and Westernization. Certainly, our friend Lino will not agree with most of the ideas supported in this book and will use them to call us "*sanfedisti*" – the name given to the bands of loyalists and counter-revolutionaries who fought against Jacobins in the kingdom of Naples to restore the authority of the House of Bourbon – a definition that, with all the affection with which he expresses it, cannot offend anyone.

If we tried to acknowledge all those who contributed to this book, the 'preface of the authors' would be more substantial than the text, so it is advisable to stop here. However, the first author cannot fail to thank again his wife, Franca – for 49 years inspirer, editor, companion and friend – who, as always, read the manuscript, performing her editing work professionally and making it readable. If she could have done the same thing to that essay more than 50 years ago, she would have made it so readable as to prevent the professor from giving me that very bad mark.

Giancarlo Genta
Torino, January 2019

In an era in which a growing percentage of the Earth's population, including this author, can be defined as "digital native", technology is now a cultural phenomenon in which can we all more or less understand each other. It is a real common language. In the same way that it happens with our written and spoken languages, technology is also used by a considerably larger number of people than the percentage who really know how it works. The vast majority of digital natives, on closer inspection, have no scientific awareness, but make daily use of devices that required an in-depth knowledge and revolutionary innovations to bring about. But above all, again just as happens with linguistics, we tend to consider technology as a fixed, consolidated and irreplaceable system.

For the latest generations, and perhaps not just for them, the temptation to consider the scientific and technological achievements of the contemporary West as ever-present, consolidated and universal is stronger. As we will see in this book, this assumption is wrong, and the risk of a drastic involution is always lurking. The aim of this book is to try to understand what lies behind this language that today seems to have become universal, as old as humankind, and the basis of any representation of its future. Behind every instrument that surrounds us, and even at the core of the same contemporary notion of "technology", there is a precise vision of the world, without which we would probably live in a very different society. This book outlines a historical path strongly focused on the West. This is not an arbitrary delimitation of the field of interest but is the result of the simple observation that technology, in the contemporary meaning of the term, is a typically Euro-American product. In fact, the archaeologists document that the development of an empirical technology has been a widespread phenomenon across the planet since the most remote times.

With the passing of the centuries, the flowering of these technologies has led to surprising results, for example, in the Indian and Chinese civilizations. But the conditions that made possible an epochal turning point, namely the advent of scientific technology, were only met in Europe. However, one could argue on closer inspection that not even science is a European exclusive. Mathematics and geometry, for example, were widespread in India, and the invention of algebra must be ascribed to the Persian civilization. This is a quite correct objection.

The real novelty, however, comes from the theoretical-practical marriage between science and technology which only occurred in Western civilization. Moreover, this had its roots in the medieval era that is often unjustly mistreated in the history books. This was a macroscopic phenomenon, the result of the encounter between the philosophical rationalism of ancient Greece and the Judeo-Christian religious concept. With this revolution, the identity of technology was radically rewritten, existing knowledge and discoveries were organically systematized through the scientific method, and the foundations were laid for their application in a systematic way on a scale that until then was unthinkable. In the West, technology became a driving force of the economy and the civilization, becoming capable of recombining new discoveries with data that had been known for centuries in a constant renewal. In this way, more and more complex devices have been created, and are bringing the possibility of projecting *homo sapiens* "beyond the horizon" of planet Earth much closer.

To some extent, speaking of technology is synonymous with talking about the West itself, to the point that it is legitimate to consider these two elements as part of a single reality. Over the last century, other cultures that have approached this type of technology have also taken on a new identity, almost automatically undergoing a process of "Westernization".

While talking about technology, this book also examines disciplines and issues that would appear to be quite separate from each other. In some chapters (Chapters 1, 2, 5, 7, 9, 10), as might be expected, we are talking essentially about technology, tools and inventions. In others (chapters 3, 4, 6, 8 and 11) we speak mainly of history, philosophy, theology and economics, and ample space is dedicated to figures like Hippocrates, Anaxagora, Plato, Jesus Christ, Saint Paul, Saint Augustine, Eriugena, Ugo di San Vittore, Luca Pacioli, Kant and Heidegger.

The contribution of two authors with different skills and backgrounds is evident even to the most casual reader. At the same time, however, the work initially carried out autonomously on the individual sections was then merged into a single discourse by both authors, since the basic thesis at the core of this book is strongly unitary and cohesive. Even those subjects and themes with a more "humanistic" origin are, in fact, an integral part of the main discourse, notwithstanding the excess of compartmentalisation that distinguishes, or perhaps afflicts, contemporary knowledge.

It is obviously unlikely that Hippocrates, Plato, Saint Paul or Eriugena ever contributed to the design, or even to the conception, of a technological device. At the same time, however, choosing to ignore their contribution to the history of human thought, and consequently the influence that the latter has exerted on the world of science and technology, has led to serious conceptual errors that have consolidated in the collective imagination with the passing of centuries, giving us a distorted idea of our own identity.

"Ideas matter", wrote Rodney Stark in the introduction to his *How the West won*. He was completely right. To some extent, this book is but a portrait of a much larger historical process, in which philosophy, religion, science, and technology have influenced one another.

Paolo Riberi
Centallo, January 2019

Preface by Giuseppe Tanzella Nitti

Today, it is a common observation that science and technology are bound to play a decisive role for the future of the human species, in the life of individuals and in their social organization, irreversibly influencing their way of life not only on planet Earth but even beyond its horizon. A horizon – it is worth remembering – that human beings were able to observe for the first time in its entirety just 50 years ago, when they ventured to lunar orbit and then reached the surface of our satellite.

Technology in particular closely accompanies our every step and gesture, makes our relationships and our movements possible, guides our learning, directs our work and our leisure, awakens our emotions, conveys our feelings. For all these reasons, a philosophical reflection on our scientific and technological progress has become necessary, a reflection that can highlight its historical roots, humanistic dimensions, ethical implications and social impact. However, precisely because of the delicacy of the issues involved, it is not easy to find balanced and objective studies on this topic. The media and popular imagination often convey biased images of science and technology, conditioned by opposing extreme views, often spoiled by prejudices and aprioristic ideas. Scientific progress is presented as a panacea that will finally give us a better world, an almost unlimited longevity, the freedom from fears and superstitions. But sometimes that same progress is depicted as the cause of catastrophic scenarios, the origin of the evils that threaten the environment and overwhelm personal life, the engine of irreversible transformations bound to change our biology, our psyche, and probably our very being, leading us inexorably towards a *post-human* stage.

If the situation is that described above, then saying *'How technology made us human'*, as in the subtitle of the Italian version of this volume, can be provocative, and somewhat intriguing. This work by Giancarlo Genta and Paolo Riberi offers us an unusual itinerary, a history of technology deliberately intertwined with the history of philosophy, trying to grasp the mutual implications of the two. Our philosophical beliefs and our visions of the world affect and determine the way we face life, our way of knowing and working. They provide us with the conceptual tools to interpret nature and to transform it, but also to alienate it or humanize it. Among these visions of the world there are also the religions of our world, in particular the Judeo-Christian tradition, which the authors acknowledge as having an important influence on the construction of Western civilization, and especially on the rise of scientific thought. In particular, it is the meeting between the latter religious tradition – the depositary of a precise philosophical view of the relationship between God and nature – and Greek philosophy centered on the rational characteristics of the *Logos*, that created the conditions suitable for the birth and growth of a mindset that allowed the scientific study of nature. The Greek culture had in itself produced some specific potentialities, also in the technical-scientific field (as Genta and Riberi remind us, appropriately recovering the works by Lucio Russo on the "forgotten revolution", such was the incipient Greek scientific thought which was then aborted or forgotten) but the dominant religious Greek culture at that time, centered on polytheism and irrational myths, did not encourage this development. The "detonator", or rather the catalyst that gave rise to science, was later provided by a religious culture where rationality and philosophy could exist together, as happened with Christianity.

In developing their arguments, Genta and Riberi are in good company with many twentieth-century scholars, from Pierre Duhem to Alfred North Whitehead, from Alexander Koyré to Stanley Jaki, from Alexander Kojève to Peter Hodgson, from Edward Grant to Alistair Crombie. The same idea had been revived, but not without controversy, by the more recent works of Rodney Stark, whose thoughts the authors of *Technology and the Growth of Human Civilization* willingly endorse. In the Greek world view, "*the general rule that governs the universe is not rationality, but the chaotic inconstancy of the gods, who do not comply to any general rule and manipulate the lives of humans as they please [...] the enterprises of humans succeed, or fail, only to the extent that the quarrelsome and manipulative gods arbitrarily decide. When the same action is repeated in the same conditions, the results may be opposite*".

The best philosophy, it is true, tried to move along more rational paths, but the general effect was that everything that was added at the philosophical level and at the level of experiential knowledge was removed from the world of the gods. The role of philosophical reason, therefore, was above all that of "de-mythologizing"; so there was little possibility, Genta and Riberi say, of a "cooperation" between the philosophical and the religious visions of the world that would allow science to be an outsider no longer. It would be the Judeo-Christian tradition to show, for the first time, that religion, philosophy and rationality could cooperate well together.

This is not the only historical clarification that Genta and Riberi provide. In this volume, we find, for example, an interesting (and against the mainstream) analysis about the liveliness of science and technology in the Middle Ages, or about the relationship between Christianity, slavery and the dignity of work in the Renaissance period. The use of mathematics and the first "modern" scientific experiences is relocated to its proper historical development, making it rooted in the university spirit of the schools of Grosseteste and Buridan, even pushing them further back to the experiences and observations by Johannes Philoponus. In their study of the roots of science and technology, the authors go beyond the commonplace and simplifications, examining the flows and refluxes of irrationality and superstition, distinguishing between what contributes to the development of authentic scientific thought and what instead hampers it.

Coming to the modern and then contemporary era, we can appreciate the balance, I would even say the courage, with which Genta and Riberi face the most controversial problems posed by scientific and technological progress. We delve into the nuclear issue, from the dramatic events of 1945 to the choices that will engage our future, evaluating documents and positions, illustrating testimonies and suggesting new outcomes. Technology also means armaments, offensive power and the possibility of destruction. It is the "dark side" of technology, which is examined carefully and without compromise, but also avoiding media and ideological drifts. From this complex situation, we cannot emerge with the easy but inadequate label of the "neutrality" of technology, a position which rightly is not suggested in this book. Instead, the authors underline the intrinsic human value, the continuity with human intelligence and ingenuity, and the positive potential of technology for the development of peoples. To endorse the neutrality of technology would in fact be an illogical and misleading operation, like trying to support the neutrality of our hands and our brain with respect to what we conceive and realize. Technology is man, and it is man as a whole, because it has contributed (also) in a certain way, to make us human. Keeping in mind the idea that Western civilization has inherited from the Judeo-Christian tradition the vision of man as "cooperator" of God, because he had been created in His image to lead to its fulfillment a world still "on its way", the authors convey a positive vision of technology and scientific progress, showing that what has been and is responsible for that "dark side" was its ideological and political enslavement, not science as such, which remains an active and constructive expression of human dignity and of the emergence of culture on nature. The authors correctly question: "*Not surprisingly, technology has also played a central role in the violent affirmation of this 'dark side' of the West, especially with the transformation of scientific achievements into weapons of mass destruction in order to maintain its rule over subjugated populations. But are we really sure that the blame must fall on these technological tools?*".

The idea that culture and science constitute a genuine promotion of the human species, the fulfillment and expression of its dignity, supports much of the analysis and considerations that the reader will find in this volume. Science is part, and an integral part, of a true humanism; a humanism that we could define as culturally secular but not atheistic, because Western culture, where science has arisen and developed, has founded the dignity of science on a vision of humankind positively called by its Creator to transform and humanize the world. Technological progress offers a decisive contribution to the development of the relational network of human relationships, a particularly important contribution to the very relational nature of human beings; it favors closeness among peoples, the mutual enrichment of different cultures, making humanity more and more a single family. Technology therefore has an important potential for the promotion of education, justice and for the reduction of social and cultural disparities. Technological progress, properly exercised, can improve the living conditions of the populations, optimize the production and distribution of food and energy resources and improve health conditions. All this becomes, in fact, a service to the integral development of the person. But "humanizing the Earth" also implies on some level the dimension of rest, of contemplation, the authors observe, that the perspective of Genesis places on the seventh day which follows the creation of man. Technology, therefore, should also favor this dimension, increasing free time, improving living conditions and allowing the development of "spiritual" activities, through a progressive control on natural and mechanical processes.

By developing what Genta and Riberi do not say but, judging from what they write, they would probably approve, one could say that without hope and without a vision of the future there could not be real scientific progress. The absurd and the nihilism – which deny a transcendent foundation at the origin of the dignity of man and his work – could in fact preserve this hope only on two conditions: to place it completely in man as a finite being, and as a useless passion, a solution unable to sustain the commitment to building society for a long time; or transforming humans into gods, thus exposing them to the disappointments and totalitarianisms that history has shown us.

In this sense, Genta and Riberi seem to distance themselves from both Richard Dawkins and Yuval Noah Harari. With regard to the British biologist, it is because they believe that technology should be seen as the fruit of freedom and not of necessity, as an expression of constructive cooperation and not of the struggle for survival. The authors do not share the idea *"that intelligence, and then technology, are nothing more than tools of our genes to expand even beyond the barriers of our planet, potentially colonizing an endless number of planets, something that, for all we know, no other living being has ever managed to do. Obviously this way of seeing ourselves and our role in the world does not satisfy us at all: the idea of being just the instruments of the will and of the lust for power – always in a figurative sense, obviously genes cannot have either a will or a lust for power – of our genes does not correspond at all to the idea we have of ourselves and is, so to speak, humiliating*". They do not even subscribe to the idea of the Israeli historian that technology has transformed us, and should transform us, into gods as a result of our cultural evolution. Technology, Genta and Riberi rightly observe, does not transform us into gods, but makes us become more human.

It remains clear, however, I add from my perspective as a theologian, that a cultural and philosophical reflection on technology that wishes to enter into dialogue with Christianity must sooner or later consider the mystery of man's moral life, the enigma of the dichotomy between the good that man feels capable of doing and the evil that his actions often unfortunately do. Technology, like any human activity, must in a certain way be "redeemed" from sin, from *hubris*, from the lust for power. Christian theology therefore points its attention to the mystery of the Easter of Jesus Christ, capable of showing that the meaning of man's original lordship over creation is, in Christ, that of service, and that the only "form" really capable of transforming scientific progress into human progress is, again in Christ, the form of charity.

Beyond philosophical or ethical convictions, the enemy to beat, to put it one way, is ignorance. It is culture, the source of true humanism, that we should really invest into.

"*The possibility of continuing on our way to a future in which the promises of technology will come about depends on the possibility of overcoming the wave of irrationalism and ignorance that seems to overwhelm us, to resume, on the basis of our roots that start from the Greek and then the Judeo-Christian civilization, the path of rationality that brought us to where we are. Above all, this is what we owe to other cultures that, in the perspective of globalization, have begun that path of modernization and Westernization that Huntington spoke about; Westernization that does not mean subjugation to the West, but simply using what has been the 'dogma' on which Western civilization was based: the world is rational, and human reason is the instrument that allows it to be understood*".

This is the wish that we formulate: to really come back to thinking and reasoning, in a scientific, deep and not superficial, humble and non-ideological way. Humility is a duty because we live in a universe that we have not created but received, a universe where we all share the fortune of having opened our eyes, a universe that increasingly, thanks to science and technology, we are surprised to understand and to transform.

Giuseppe Tanzella-Nitti

Pontificia Università della Santa Croce, Rome
Professor of Fundamental Theology
Vatican Observatory

December 2018

Authors' Introduction

Technology is as old as the human species. Since it appeared on this planet, *homo sapiens* has employed objects they have transformed with their hands to make them suitable for a wide variety of uses. The relationship of the human species with technology is complex. On the one hand, and this is obvious, humans have produced a constantly evolving technology. On the other hand, technology has deeply transformed human beings, influencing first their physical and then their cultural evolution. The role of technology in human history is controversial. While many see in technology a force that has allowed us to overcome, at least up to a certain point, the limitations of our human nature, others conceive it as a force that, by causing us to depend increasingly on those machines we have created to solve our problems, dehumanizes us and makes us similar to them.

Our world is an utterly artificial one, created by humans and for humans by deeply transforming nature. We are unable even to think about how our life could be in a pristine world, not transformed by our own hand, and those of us who dream of a return to nature do not realize that what they dream of is nothing but 'old technology', the technology of a few centuries ago (or maybe just decades ago, given the current speed of progress).

Actually, it has been since at least the Neolithic Revolution that humankind has lived in an increasingly artificial world. As Shakespeare said centuries ago, a rose or a wheat field did not exist in nature; they are entirely artificial things, the product of human ingenuity.

A field of wheat is thus no more natural than a nuclear power plant or a spacecraft, and if we do not realize that, it is because we are so used to its presence that we are unable to imagine a world in which wheat does not exist.

It is technology that makes us human, allowing us to overcome the limitations imposed on us by nature. If we have problems with that, it is most likely due to the acceleration of technological development, which, owing to its speed, creates difficulties in adapting to a world in continuous transformation.

However, it is not at all said that this trend toward increasingly faster progress should be taken for granted. On this matter, two schools of thought exist.

On the one hand, some believe that this acceleration will continue at a growing rate and speculate that, at a certain point, there will be a 'singularity', a point in which technological progress will accelerate beyond the possibility, for modern human beings, to understand and foresee it. Some futurologists predict that this singularity will occur within 10 to 30 years.

On the other hand, there are those who believe that progress, first scientific and then technological, is already slowing down and that, in the not too distant future, it will stop completely, creating a situation of stagnation. Recent phenomena – the so-called 'scientific illiteracy' of large sections of the population, the multiplicity of irrational beliefs and pseudo-sciences, the widespread denial of science and technology – make one concerned that this scenario is more than just a groundless fear.

This short book, written to be accessible to readers with a non-technical or non-scientific background, aims to enter this debate on the importance of technology for human development and on the possible future of our civilization.

Technology dates back to prehistory when, even before becoming fully self-conscious with the so-called cognitive revolution, human beings began to make tools and to control fire. At the end of prehistory, with the Neolithic Revolution, they learned to change the environment in which they lived radically, creating new species of animals and plants. This phase of human evolution is the object of the first chapter.

The next chapter describes those changes which led humankind to move from prehistory to history, analyzing in greater detail the main aspects of this transformation; the invention of writing and money and the spread of slavery

The third and fourth chapters deal with one of the many civilizations which flourished on our planet, the Greek civilization, which based its knowledge of the world on the rational study of the universe and of ourselves. The encounter between the rational approach of Greek philosophy and the Jewish religion, which had extended this rationality from the physical to the metaphysical world, is described, to move on to the birth of Christianity and to that synthesis, also based on rationality, which is the Greek-Judeo-Christian world.

The Greek-Judeo-Christian civilization was spread across the Western world by the Roman civilization which, despite the markedly anti-technological ideology of its ruling elite, ended up spreading technology it inherited from the previous civilizations, and the rationality of Greek and then Jewish Christian origin. This is the topic of Chapter 5.

With the fall of the Western Roman Empire, the Western world found itself in a period that is often considered to be backward and barbaric. In Chapter 6, we discuss how this vision of the Middle Ages is tainted by misunderstandings and prejudices, and how, on the contrary, it was precisely in this period that, under the influence of Judeo-Christian rationality, technology began to take that step which brought us from the empirical technology of the ancient world to the scientific technology of the modern world.

This passage, started in the Middle Ages, could only be completed with the development of modern science. Chapter 7 describes, as examples, some technologies related to steam engines, to the creation of flying machines, electromagnetism and finally nuclear energy; technologies that, by their very nature, could not be developed empirically through the trial and error approach, but needed a scientific knowledge of nature. Finally, a paragraph is dedicated to the birth of engineering schools, which trained the men (and more recently, the women) who made it possible to re-establish technology on a scientific basis.

The shift from empirical to scientific technology, with the quick technological progress that followed, did not happen all over the world, but was a feature of Western civilization, which found itself enjoying considerable advantages over the rest of humanity. Technological development was accompanied by a change in the economic structure of society in an increasingly capitalistic sense and the result has been an imperialistic expansion of the Western powers. This process has not been painless and has produced some of the worst pages in the history of Western civilization. Two paragraphs of Chapter 8 are devoted to the worst aspects of this process.

The same technology was involved in these negative aspects of the imperialist expansion, bringing the destructive capacity of humans to levels never achieved before, at least in a relative sense, as the destructive capacity of every single agent employed in destructive actions. Chapter 9 is dedicated to this "dark side" of technology.

Chapter 10 deals with what is normally called the Industrial Revolution, distinguishing its various phases and trying to underline their differences and common characteristics.

Parallel to this effort, aimed at understanding the world through the rationality of Greek-Judeo-Christian origin, an irrationalistic current has developed throughout the history of humanity, one that tries to interpret the universe through a magic-esoteric approach. This current is not at all confined to the past and bound to disappear; on the contrary, it has also recently hidden itself among the most dogmatic scientific and rationalistic tendencies and, in the last century, has strongly impregnated Western philosophy itself. Chapter 11 is dedicated to this irrationalist constant, trying to understand whether this irrationalism, which seems to have spread more and more, may lead to the decadence of the West by aborting the path of ever-increasing development based on a rational understanding the world.

Finally, Chapter 12 tries to understand where technology can take us – particularly in view of an expansion of the human species into space – if the problems analyzed in the previous chapters can be solved by the same Western civilization or, in case it abdicates its proactive role, by other civilizations that could take its place.

A short epilogue, in which we try to assess the advantages that technological progress has brought to humanity, concludes the text.

1

Technology in prehistory

1.1 A TECHNOLOGICAL ANIMAL

Man is a "technological animal". This is one of the many definitions which were proposed in an attempt to outline the difference between humans and the other animals. Actually, one of the basic characteristics of the human species, and partially also of the hominids who preceded it, is its ability to develop the technology that, from the early advances of the Stone Age, has today led us to leave our planet on our first, hesitant attempts to travel through space.

From this point of view, humans are radically different from all the other animals which have populated, and still inhabit, our planet. Since life appeared on Earth – on the possible presence of living beings elsewhere in the Universe we have no certainty, and therefore we cannot comment – a huge variety of living beings have developed, and each species has adapted to the various environments through a process, still not completely understood, based on mutations and natural selection. This process allowed a range of living species to specialize for the various environments, so that life could colonize the entire planet.

With humans, something radically different occurred. For the first time a 'generalist' species developed, one that was able to survive and then multiply until it became dominant in a wide variety of environments, without specializing for any of them.

For example, animals developed a thick fur, constituting a thermally insulating layer, to adapt to cold climates. Humankind instead learned to use animal skins to make clothes which perform a similar function. Over time, humans learned to spin and weave fibers obtained from plants or animals – like cotton and wool – creating materials which do not exist in nature. Eventually, humans created synthetic fibers to produce textiles whose raw material did not exist before the related technological processes were developed.

Similar processes allowed this species, lacking the physical characteristics required to play the role of predator (fangs, claws, etc.), to invent the weapons and tools to allow it to capture its prey and subsequently to separate the edible parts from those which must be discarded (the separation of meat from bones is practically impossible using only the 'physiological' tools of the human species). Later, other inventions made it possible to transform these waste materials (bones, skin) into other tools and objects useful for a large variety of purposes.

Actually, it is not entirely accurate to say that humans are not specialized. It is simply that their specialization is of a different kind. The specialization of humankind is thought, and the organ which permits this is the brain.

The tools and other artificial objects that humans have invented have been used as prosthetics to be added to their bodies. They allowed humankind to play a large number of roles and to colonize different environments. From this point of view, humans were defined as "*added monkeys*" [1].

The biological evolution of the human species took place over a number of steps, all with regard to their nature as generalist animals.

If we look at the genealogical tree of modern humans (Fig. 1.1) since the beginning of the Oligocene (37 million years ago), we can see how the apes branched off from the monkeys, from which the various hominid primates later developed. After the separation of the gibbons branch, about 10 million years ago, the differentiation between the pongids (orang-utans, gorillas and chimpanzees) and the hominids took place, probably with the birth of the *australopithecus*. This evolutionary line led to the appearance of species with larger brains (Fig. 1.2), even of dimensions never reached before (at least for the ratio between the weight of the brain and that of the body)

The main line, leading to the development of the human species, was therefore related to the development of the brain. Actually, the transition from apes to humans happened gradually, and several species of hominids with increasingly larger brains preceded modern humans (*homo sapiens*) over a span of almost three million years[1]. However, it is not only a question of brain mass, but also of structure. While evolution led to ever-increasing brain size, the cerebral cortex became

---

[1] In this work, modern man will be defined as *homo sapiens*, instead of *homo sapiens sapiens* as was customary in the past. In the same way, instead of *homo sapiens praesapiens* and *homo sapiens neanderthaliensis*, the terms *archaic homo sapiens*, or *homo heidelbergensis*, and *homo neanderthaliensis* will be used. The most ancient *homo sapiens* is the Cro-Magnon man.

1.1 A technological animal

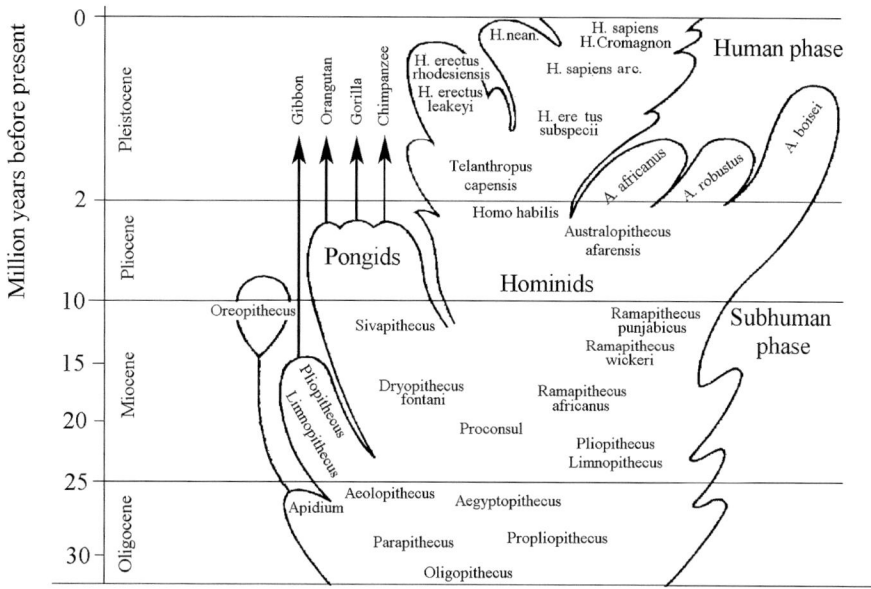

Figure 1.1. The Hominoidea. Evolution of man and apes starting from common ancestors in the Oligocene, about 30 million years ago. Note that the timescale is strongly nonlinear (it is much expanded towards the present). [Image modified from G. Genta, *Lonely minds in the universe,* Springer-Copernicus Books, New York 2007.]

more complex with a large number of convolutions, permitting an increase in the brain surface within the same total volume.

Together with this trend, other evolutionary lines modified the anatomy of the pharynx, allowing it to produce a greater variety of sounds. This line was instrumental to the development of language, and it occurred later than the development of the brain. Neanderthal man (*homo neanderthaliensis*), which appeared on the scene in relatively recent times and almost simultaneously to *homo sapiens*, had a much more primitive larynx and likely was unable to develop a true language. This was probably an extremely serious handicap, owing to the importance of language in the development of the ability to think.

A third evolutionary line, fundamental in human development, was the one leading to the ability to walk in an upright position. It is not so much a matter linked to modifications in our bones – actually, we are not yet fully adapted to this posture, as evidenced by the frequent spinal pathologies to which we are subjected because we use our spine in a way that is quite different to that for which it developed – as much as to the development of the organs of equilibrium in our inner ear. Our ability to detect accelerations precisely – which is unique in the

4 Technology in prehistory

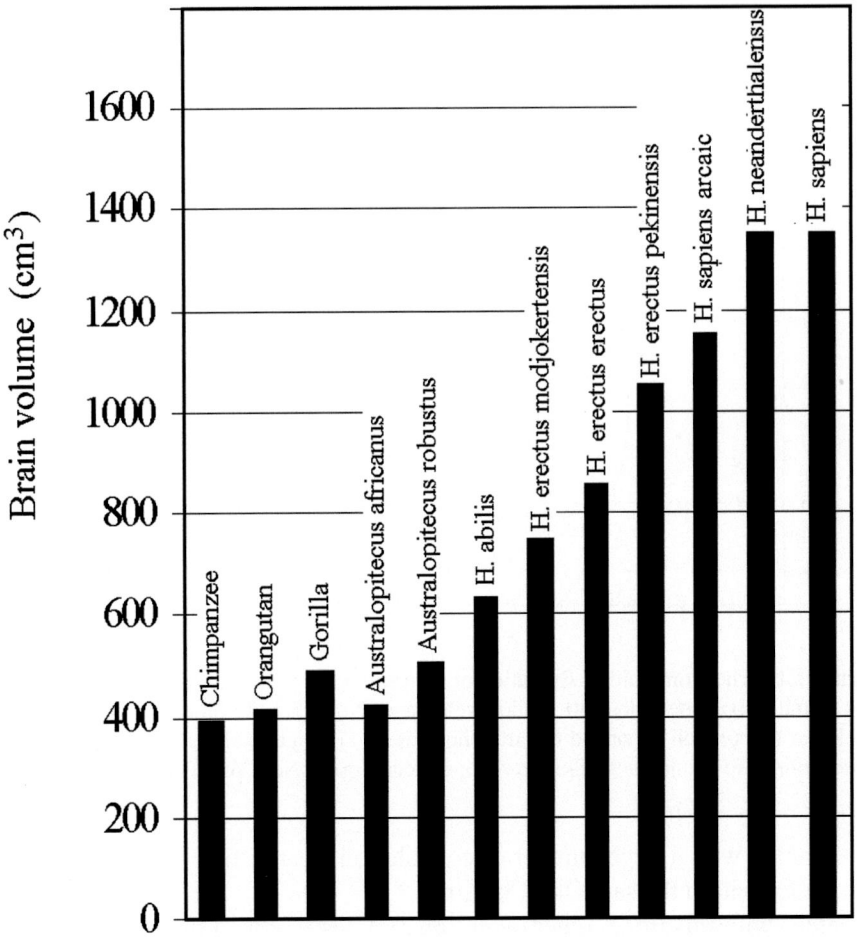

Figure 1.2. Average volume of the brain in apes, hominids and humans. [Image modified from G. Genta, *Lonely minds in the universe*, Springer-Copernicus Books, New York 2007.]

animal world – allowed us to maintain an upright position, but later proved to be essential for driving motor vehicles and flying aircraft. Walking in an upright position allowed humans to transform the front legs (organs devoted to locomotion) into arms (organs devoted to manipulating objects), an essential step for the development of technology.

Taken together, the changes dealing mainly with the brain, the larynx and the equilibrium organs (as well as the hands) are what distinguishes us from animals anatomically.

1.2 STONE IMPLEMENTS

The use of objects to perform a number of tasks is not exclusive to humans, and some animals may occasionally use tools. For instance, fur seals sometimes use a pebble to break the shell of mollusks, while some monkeys use a piece of wood as a club, or a twig to take the ants from an ant hill.

Certainly, even the earliest hominids occasionally used pebbles with reasonably sharp edges, discarding them after use. However, only humans are able to build a tool with a view to its future use and to keep it after using it to perform intended actions.

The oldest purpose-built stone tools date back as far as 2.5 million years ago. Conventionally, we speak of the Lower Paleolithic to designate a period of time ranging from about 2.5 million to 120,000 years ago, a period in which *homo habilis* and *homo erectus* spread across Africa, Europe and Asia.

We did not have to wait for the development of modern humans see the arrival of stone tools, however. The limited intellectual power of *homo habilis* was still sufficient to develop the so-called Olduvai culture (although the name comes from the town of Olduwai in Tanzania, this culture spread into Europe and Asia as well). In particular, paleontologists believe that the hominids of that period had not yet developed either a language or full self-awareness and therefore they cannot be considered as intelligent and self-conscious beings. Nevertheless, they were able to make tools anticipating their future use and to keep them after that use, a feat which, as already mentioned, no animal is able to perform.

The tools of this culture were typically just chipped pebbles which could have a fairly wide range of uses. Regardless, their production required a good practical knowledge of the material which, evidently, slowly consolidated over millennia.

About 750,000 years ago a new human species, *homo erectus*, developed a new culture, the so-called Acheulean culture (from the site of Saint-Acheul, near Amiens, in France). This culture is characterized by stone artefacts having the shape of almonds, worked on both sides in a symmetrical way ('bi-faces' or 'amigdals'), and is often associated with different tools made from splinters (scrapers and points).

Conventionally, the transition from the Lower Paleolithic to the Middle Paleolithic is set about 120,000 years ago (a tentative chronology of prehistory is shown in Fig. 1.3) and coincides with the appearance of a new human species, *archaic homo sapiens*. The culture, dating back to the period from 120,000 to about 36,000 years ago, is normally defined as Musterian, from the site of Le Moustier, in Dordogne, France. The artefacts of that period are much more varied and are made from different materials such as wood and leather, as well as stone. Furthermore, there is a greater differentiation between the various sites, a characteristic that shows a diversity between the various cultures. The behavior of *homo*

6   Technology in prehistory

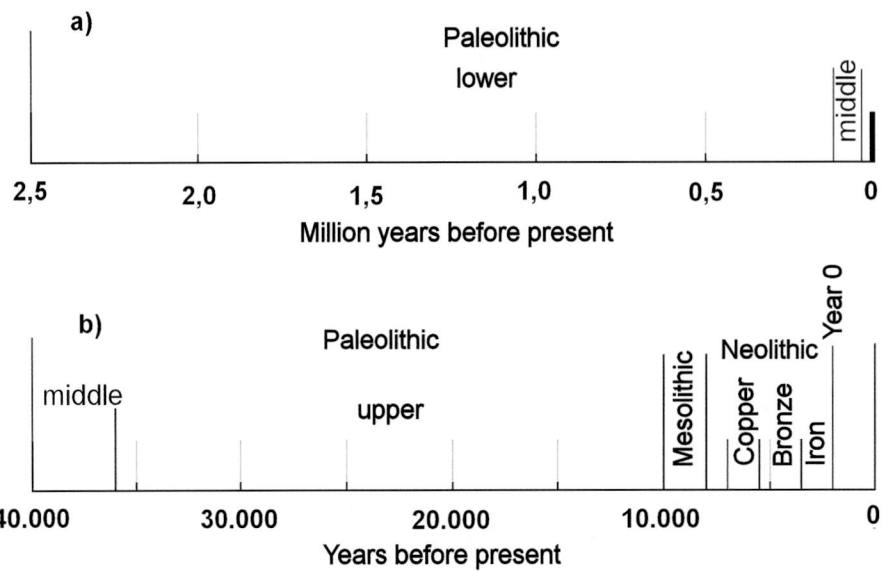

Figure 1.3. Tentative chronology of prehistory. a): 2.5 million years before the present. For the main part of this period, humans lived in the Lower Paleolithic. b): Last 40,000 years before the present. During this period, the Mesolithic, the Upper Paleolithic and the Neolithic were developed. Only since the last millennia before year 0 did humankind emerge from prehistory to enter actual history. However, it should be noted explicitly that, starting from the Mesolithic, the chronology differs from place to place and, in certain areas of Earth prehistory, lasted until relatively recent times. The terms 'Copper, Bronze, or Iron Age' (see below) have very little meaning in an absolute sense, (as their chronology is extremely variable), but are useful in relative terms to indicate the different stages of development of the various civilizations at different times. The chronology shown here refers to the first civilizations that went through these stages.

sapiens, and that of its contemporary *homo neanderthaliensis*, is somewhat different from that of its predecessors. In particular, among the characteristics of the middle Paleolithic there are:

- Hunting of large herbivores by attracting them into natural traps,
- Transportation of flint over distances of hundreds of kilometers,
- Construction of huts,
- Burials, accompanied by ritual objects,
- Collection of ocher and incisions (non-figurative) with presumably aesthetic purposes,
- Probable invention of the early musical instruments, similar to flutes.

1.3 THE COGNITIVE REVOLUTION

The last three points seem to indicate a radical change. For the first time, we can see the presence of self-conscious beings on our planet, who can 'think' and ask themselves questions going beyond their immediate needs, although humans at the beginning of the Middle Paleolithic were likely not yet fully self-conscious. The Middle Paleolithic is a period of transition towards full self-consciousness, which would manifest itself only in the Upper Paleolithic with *homo sapiens* and probably with *homo neanderthaliensis*.

An interesting hypothesis is that described by Yuval Noah Harari in *Sapiens. From animals into gods, a brief history of humankind* [2]. In the second half of the Middle Paleolithic, between 70,000 and 36,000 years ago, an actual revolution broke out. It did not affect technologies or the material conditions of life but was triggered by the development of new abstraction capabilities of the human mind. Humans, in particular those of the species *homo sapiens* but likely not those of the Neanderthal species, began to think about things beyond their direct sensorial experience – in reality, Harari says, things which do not exist – and to believe that these things had a real existence next to those belonging to the material world in which they lived.

Following this theory, these objects that go beyond the concrete world would have been created by humans gradually, starting from the cognitive revolution with a process that continues, perhaps at an increasing pace, even today.

Among these non-existent objects, Harari includes not only the entities described by all the religions of humanity, from the spirits of the most primitive religions to the gods of polytheistic religions, to the only god of monotheistic beliefs, but also the values of which our political doctrines speak – freedom, equality, human rights, etc. – economic and legal entities from money to corporations and, we could add, the entities that have been gradually introduced by natural philosophy and then by science, such as forces, work, and power. Speaking of science, we could include here all the things which are too small – or too large – to be detectable without adequate equipment, such as atoms (from the birth of the philosophy of Democritus to modern atomic physics), galaxies, or entities that do not yet have any experimental evidence like strings. It could be concluded that this creation of non-existent entities has reached its peak with modern physics, on the one hand with the theory of relativity and on the other with quantum physics.

In reality, the point is not so much the existence of these entities – hardly any physicist would define forces as non-existent, nor would an economist do so regarding corporations or money, and the faithful of any religion would never deem their divinity as non-existent – but the ability of humans to behave as if entities that are not detectable by the senses actually belong to the world surrounding us. The cognitive revolution consists of the ability of abstract thought, which probably stems from the availability of a more articulate and flexible language and the continuous exchange of opinions among the members of the human community. Not only have none of the various animal species, not even the most evolved apes, reached this stage, but neither did the other human species.

The doubt remains that *homo neanderthaliensis* reached this stage along with *homo sapiens*. On the one hand there are favorable indications linked to archaeological findings which suggest that they, too, had religious beliefs and developed artistic behavior, but there are also negative indications, in particular related to the different configuration of the larynx which would render this species incapable of sufficiently evolving a language. Perhaps the truth lies in the middle. Recently, it has become more certain that a degree of inter-breeding between the two species was possible, as often happens in the case of species not completely separated. In this case, there would have been intermediate individuals and perhaps even groups between the two species which could have participated in the cognitive revolution and then, at least in part, were absorbed into the dominant species. Perhaps we owe to these groups the small contribution of the Neanderthal genome that many of us still possess today.

From the viewpoint of technology, all techniques related to stone chipping progressed at an accelerated pace and the cultures diversified. Unfortunately, the use of less durable materials, such as bone and hide, makes it more difficult to document these advances archaeologically.

There is little doubt that the differentiation among cultures was the most significant outcome of the new importance of the ability to think in abstract concepts, in a life which could no longer be defined only in terms of the material.

The Upper Paleolithic, which spans from 36,000 to about 11,000 or 10,000 years ago, corresponds to the final stage of the physical development of man, with the birth of the species to which we also belong, *homo sapiens*, and that of the Neanderthal which quickly became extinct. In this period, there is the certainty of the presence of actual humans in the full sense of the term, and artistic expression became a central fact of culture. Implements made of stone, bone and other materials ceased to be shaped only in terms of their use, and started to become decorated, sometimes in a truly valuable way. Furthermore, tools and artistic objects took on ever-greater variety of forms and styles, which depended on the cultural environment. The various cultures differentiated, inventing different styles and technologies. It should be noted that this period is characterized by a particularly harsh climate (the Würm glacial period).

Technological development proceeded in a very discontinuous way during the Paleolithic. The variety and quality of stone tools underwent rapid changes, followed by periods of stagnation lasting hundreds of thousands of years (Fig. 1.4). Moreover, these quick changes coincide with the evolutionary stages that led to the development of new species of the *homo* genus. It seems therefore that the members of a certain species reached the highest level they were able to achieve in a short time (obviously, the adjective 'short' still refers to very long evolutionary periods) and then continued to repeat the same technology until a new

1.3 The cognitive revolution

evolutionary step was attained. Even the fact that objects produced by a certain species had very few local variations seems to indicate that technological ability was influenced much less by culture than by human physical structure. If such a comparison has any meaning, it is somewhat like rigid automation, where automatic machines always produce the same object following the machine's own structure (hardware), as opposed to flexible automation in which each machine can produce a variety of objects depending on its programming (software).

When *homo sapiens* and *homo neanderthaliensis* finally appeared, there was a revolution giving rise to the culture of the Upper Paleolithic.

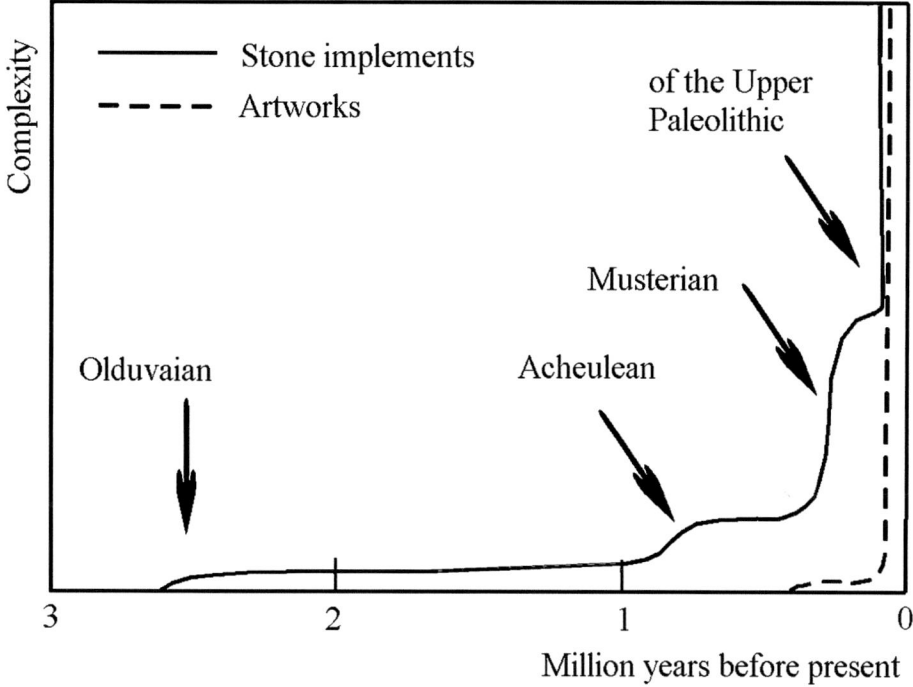

Figure 1.4. Increasing complexity and variety of stone tools and complexity of artworks in the development of humankind. Note the discontinuities linked with the appearance of *homo abilis* (Olduvaian culture), *homo erectus* (Acheulean culture), *archaic homo sapiens* (Musterian culture), and finally of *homo sapiens* and *homo neanderthaliensis* (culture of the Upper Paleolithic). [Image modified from G. Genta, *Lonely minds in the universe*, Springer-Copernicus Books, New York, 2007).]

After the cognitive revolution, humans learned how to sail across wide stretches of sea and completed their dispersal across all the dry lands of the planet. This was also possible thanks to periods when the sea level was lower than today due to the ice ages and to bridges of land which joined some continents, at least partially. The colonization of Australia took place about 40,000 years ago, while that of the Americas, through the Bering Strait, occurred less than 20,000 years ago.

At this point, the human species was living practically on the whole planet, except for some of the more remote islands and particularly cold territories. Reaching and colonizing these places would require the development of more advanced technologies.

The noosphere now became an addition to the lithosphere (the solid Earth's crust, formed mainly by silicates), to the hydrosphere (all the waters of the seas and rivers), to the atmosphere (the gaseous envelope that surrounds the planet) and to the biosphere (all the living beings that are on Earth), which had begun to form gradually since the appearance of life, probably almost four billion years ago. The term 'noosphere' was introduced by Édouard Le Roy, Pierre Tehilard de Chardin and Vladimir Ivanovič Vernadskij [3]. It indicates the totality of the thinking beings on the planet, or the collective mental activity of *homo sapiens* [4]. Actually, the meaning of noosphere is quite variable depending on the scholar who uses it.

To the noosphere was then added the technosphere, which is generally defined as the totality of all man-made objects, sometimes including even the plants and animals that have been domesticated since the Neolithic Revolution.

1.4   FIRE CONTROL

In the Lower Paleolithic, a further technological revolution took place in addition to the improvement of stone chipping techniques. Humans learned how to use fire, how to control it and finally how to produce it. It is very difficult to assess when this took place, since fire could occasionally be produced by lightning or volcanic eruptions.

There is evidence that occasional use of fire took place about 1.5 million years ago, which means that the first user of fire was *homo habilis* of the Olduvaian culture. Some evidence comes from traces found mainly in East Africa. There is evidence of fire control and of its more frequent use, through the construction of hearths, in the Acheulean culture, particularly in South Africa and in Zambia, and then in almost all the areas settled by humankind.

The use of fire was a fundamental step in the development of humans, as shown by the myths dealing with fire which are present in almost all civilizations. Almost everywhere, fire is considered as a gift from the gods, or from a hero who stole it from the gods. In our culture, the best known is the Greek myth of Prometheus.

One of the first uses of fire was in applying heat to help split stones that were particularly difficult to work with. However, fire enabled enormous changes in human life, changes that go far beyond those related to the construction of stone tools, and deeply influenced human physical evolution.

Of particular importance are the improvements in nutrition made possible by cooking foods which were otherwise difficult to digest or completely inedible. Many components of plants, such as raw cellulose and starch, as well as certain parts such as stems, leaves, roots and tubers, could not be digested raw and could not be eaten prior to the use of fire. These quickly became part of the human diet. In addition, some seeds and tubers containing toxic substances could also be made edible by cooking, while cooking made meat easier to eat and to digest, improving nutrition with its protein intake. The amount of energy required to digest cooked meat is much less than that needed for raw meat, while the heat kills off pests and bacteria that may be contaminating the food.

The use of fire was also influenced by the development of intelligence and was in turn one of the factors that accelerated the increase in brain mass, the consequence of which is even greater intelligence. Since a large brain requires large amounts of readily assimilated energy (although it comprises only two percent of the body's mass, the human brain requires about 20 percent of the energy), it could not have been developed without a meat-based diet and, most likely, without the nutritional improvement directly resulting from the use of fire.

The use and later control of fire were unavoidable steps along the way that led to the development of an intelligent and self-conscious species. However, in order to truly dominate fire, humans had to master the technology to ignite it at will and not just preserve those fires which could occasionally ignite spontaneously. For this task, the most ancient methods were likely those based on rubbing (Fig. 1.5), but as they were implemented by using highly perishable materials such as small pieces of wood, they left hardly any archaeological evidence. However, it is believed that *homo erectus* was the first to ignite fire, between 500,000 and 200,000 years ago. Much later, *homo sapiens* used techniques based on percussion, which do leave abundant traces, particularly if flint is used.

With fire, humans also had to deal with a dangerous technology for the first time. In spite of all its advantages, fire is extremely dangerous, and very small actions or trivial errors can trigger devastating consequences. This was true even in the Paleolithic, when humans possessed much fewer flammable objects and did not live in cities with wooden buildings, with enclosed spaces that could be saturated with toxic gases and smoke.

However, perhaps the most important consequences of fire control are related to social life. The light produced by fire allowed humans to increase the number of hours of activity and, at the same time, provided protection against predators of all kinds. Consider that herbivores spend almost all their time feeding and apes spend more than eight hours a day eating. The use of fire, and the possibility of eating cooked meat, allowed humans to dedicate much less time to this activity, increasing the time which could be devoted to social life, artistic activities and the construction of tools.

There is no doubt that the possibility of being together in a safe and lighted place during the long evening hours, when no external occupation was possible, allowed the development of artistic and relational activities which, thanks to language, were a consequence and, at the same time, a cause of the development of an ever-increasing intelligence.

12   **Technology in prehistory**

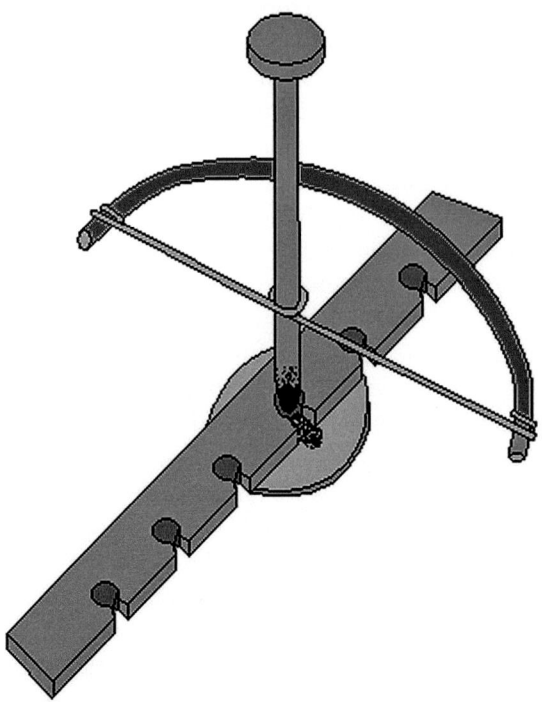

Figure 1.5. Bow drill used to start a fire (image from https://upload.wikimedia.org/wikipedia/commons/6/6f/Bow_Drill.png).

1.5   THE MESOLITHIC CRISIS

At the end of the Upper Paleolithic, humans had become very talented hunters and enjoyed a standard of living that had never been achieved before. The invention of the bow dates back to the Middle Paleolithic, probably as an evolution of the bow drill used to light fire. With the bow, humans managed to extend the range of their weapons, increasing the productivity of hunting, at least as far as small prey was concerned.

During the Upper Paleolithic, and in particular towards the end, a new weapon was introduced, the spear thrower (Fig. 1.6). While the bow is an energy accumulator, in which muscular energy is first stored and then discharged through the arrow, with the spear thrower the force is directly provided by muscles driving the launching mechanism. It required much more physical strength and greater skill but allowed the hunter to launch heavier weapons with higher speed.

With the spear thrower, it became easier and less dangerous to hunt large animals. However, the increased efficiency of hunting which took place at the end of the Upper Paleolithic was due not so much to the introduction of new weapons as

1.5 The Mesolithic Crisis   13

Figure 1.6. The spear thrower, introduced at the end of the Upper Paleolithic, made hunting for large animals much more effective. [Image from https://upload.wikimedia.org/wikipedia/commons/e/e9/Propulseur-2.jpg.]

to new forms of organization. In particular, the practice of driving herds of herbivores towards steep slopes or cliffs developed, even those of very large size like mammoths and bison, to recover their carcasses after they were killed by the fall. For this purpose, fire control could be extremely useful.

The abundance of food, as a result of hunting and harvesting wild plants and the extensive use of fire, enabled a remarkable growth of the population in all the lands where humans had settled – practically the whole planet, since they had also settled the Americas and Australia over 15,000 years ago. Humankind could enjoy the usual fruits of abundance, in particular a greater amount of free time, which enabled them to develop the art forms which had already been introduced tens of thousands of years before, like rock painting and the decoration of tools and other artifacts.

The hunting techniques developed at the end of the Upper Paleolithic enabled a level of slaughter which far exceeded the immediate needs of the population. In many sites, evidence has been found that large quantities of animal carcasses were left to rot without anyone using that food; perhaps the oldest example of large-scale food wastage.

Over a few thousand years, the number of large mammals decreased greatly and many of them became extinct. As also evidenced by rock paintings, humans went back to hunting small animals and the bow again became the weapon of choice over the spear thrower.

However, things did not develop in the same way all over the planet. In Africa, Asia and even in Europe, the continents where humans evolved, the animals had slowly adapted to their presence and had learned to fear them. In the Americas and Oceania, where humans first arrived at a much more advanced stage of their evolution, the animals came across this new predator suddenly and did not have time to adapt to the new situation. The new hunting techniques thus caused the extinction of a greater number of species.

Primarily, albeit with considerable differences between geographical areas and perhaps also due to the temperature increase at the end of the Ice Age, the number of inhabitants exceeded the acceptable value that could be sustained by the available resources. Excessive hunting led to a rapid decrease in resources which humans had at their disposal, causing a reduction in the level of well-being and an actual overpopulation crisis. It was an unprecedented event.

Fishing evolved on the coasts and near the big rivers, with the introduction of primitive floats and later of boats, but many scholars speak of an actual social and environmental catastrophe, and of a widespread setback for the human species. The archaeological remains show that, in many areas, humans had to feed on very small prey such as snails and other mollusks in order to survive, and that wars for the possession of food broke out.

1.6 THE NEOLITHIC REVOLUTION

At the end of the Mesolithic period, humans began cultivating a few plants and taming some animals – probably the first animal to be tamed was the dog. It is possible that agriculture was, at least in part, a response to the Mesolithic Crisis and allowed the human species to resume its path towards greater development that had seemed to be utterly compromised. By that time, humans had settled virtually every continent and had seen off the last direct competitor, Neanderthal man. *Homo sapiens* thus began a new phase, normally referred to as the Neolithic Revolution. Slowly, humans learned how to cultivate more plants and to breed the animal species they were feeding on, instead of gathering the wild fruits of the land and hunting the wild animals.

Living beings can be divided into autotrophs, or beings which themselves produce the organic substances they need from non-living matter, and heterotrophs, or beings that use complex organic substances produced by other living beings. The former are at the base of the food chain, while the latter, such as humans as intelligent beings which require large amounts of food to support a large brain, tend to be at the top of the food chain, even if in turn they can be eaten by other predators. With the Neolithic Revolution, humans learned to produce the organic substances they needed and, although remaining heterotrophic, transformed into a sort of technological autotrophic creature.

The availability of food thus radically multiplied, allowing a much larger number of individuals to live in a territory that could previously support just a few. Moreover, the Neolithic Revolution did not take place simultaneously all over the planet and had different results across the various geographical areas, to such an extent that it is even possible to trace back the different development of the various populations and the occurrence of recent phenomena, such as colonialism and underdevelopment, to the mismatch between the times and the ways in which it happened. As Jared Diamond pointed out, these differences are not related to some difference in the intelligence or character of the various ethnic groups, but to actual geographical differences between the areas in which they live and to differences in the animal and plant species of the various continents [5].

The main centers of the Neolithic Revolution are shown in Figure 1.7, together with the most important plants domesticated and the approximate age at which the revolution took place.

The Neolithic Revolution is, however, much more than the ability to produce food instead of collecting or hunting it. The domesticated species were gradually modified from the wild ones they were derived from, acquiring characteristics which no longer suited their own needs but those of the species which domesticated them. This created a sort of protected environment, dominated by artificial selection, in which the usual rules of natural selection no longer applied. Owing to this process, humankind started living in an increasingly artificial world, dominated by technology.

One of the best examples is the almond tree. The fruit of the wild almond contains amygdalin, a very poisonous glucoside. After the taming of the almond tree, its fruits became edible, containing smaller and smaller quantities of amygdalin. Jared Diamond believes that the decrease of this glucoside started with a genetic mutation and that the edible variety was selected by the ancient farmers in the process of domestication. Almonds were probably tamed at the beginning of the Bronze Age.

Animals not only provide proteins but also perform other fundamental functions for the development of a technological civilization, from the supply of energy to the production of fertilizers.

The Neolithic Revolution multiplied the availability of food and, in addition to solving the problems which broke out during the Mesolithic Crisis, allowed humans to restart the trend toward progress. But this was not free from risks. Humans became increasingly dependent on domesticated species and the consequences of unpredictable events, such as droughts and natural disasters of various kinds, proved to be increasingly serious. A mechanism similar to that which had caused the previous crisis re-emerged: the increase in food availability led to a population increase and made humankind dependent on its own technology. In particular, in many areas it was essential to regulate rivers and grant soil fertility. Sedentarization and living in settlements of increasing size also favored the onset of epidemics, with periodic mass deaths. In time, the populations living in urban areas acquired a growing immunity to the most common diseases, which mainly plagued populations living in less crowded areas. This mechanism repeated until relatively recent times, when Europeans came into contact with colonized populations. The sedentarization and division of labor, resulting from the Neolithic Revolution, triggered social dynamics that led to the stratification of the population into classes and sometimes into castes.

After the Neolithic Revolution started, the human species experienced a phase of fast technological progress, a progress that was subjected to an impressive acceleration from the beginning. In the Paleolithic, cultures lasted tens or

16 **Technology in prehistory**

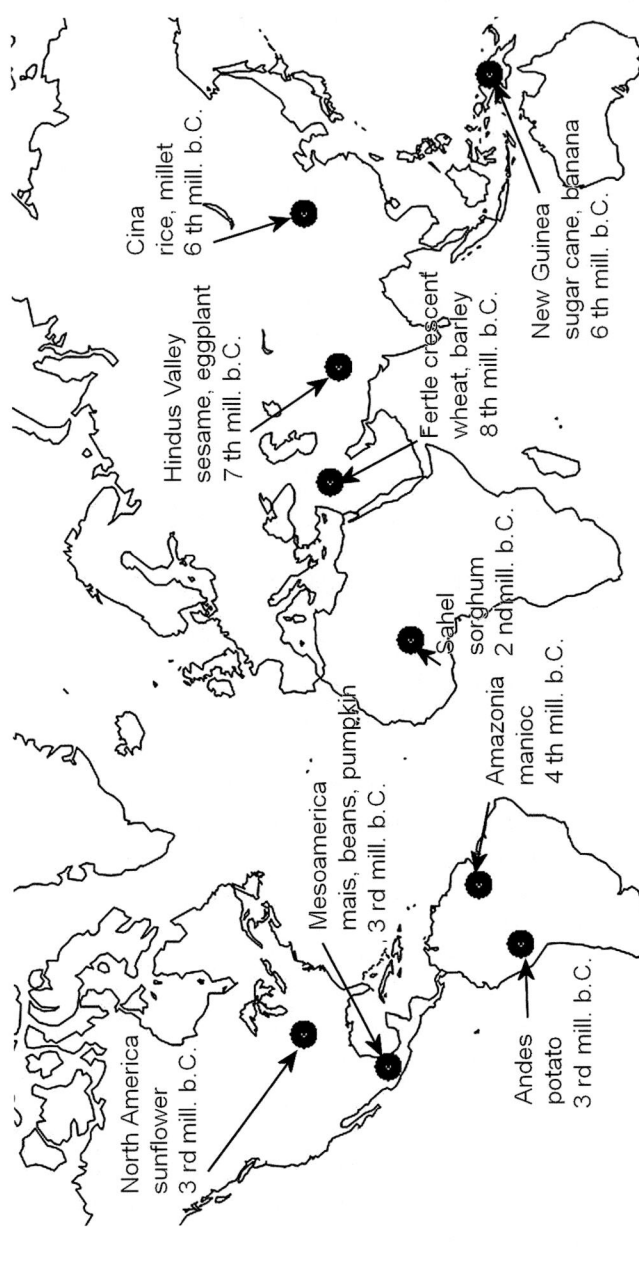

Figure 1.7. The main centers of the Neolithic revolution. The main domesticated plants and the approximate age at which the revolution took place are also shown. Dates are partially taken from H.J. Morowitz, *The emergence of everything*, Oxford University Press, Oxford 2002.

hundreds of millennia, during which there was apparently no change in the core of material life. For very long periods, humans chipped stones, always in the same way, to obtain tools which were very similar to each other. In contrast, once they started cultivating, the characteristic length of time of progress shortened, first being of the order of millennia and then of centuries after they learned to work metals. Subsequently, with the accumulation of technological knowledge and with the possibility of diverting an increasing number of people from activities directly related to the production of food, these characteristic times shortened again.

Among the technologies that the various civilizations developed during the Neolithic period (even if their development had often started earlier), a leading role is played by ceramics and weaving. These technologies had an important role in everyday life and, in turn, led to developments in other fields. For instance, ceramics required mastery of the technology of baking in the oven at temperatures higher than those needed for cooking, and this was certainly very important for the development of metallurgy.

1.7 THE AGE OF METALS

Traditionally, we speak of the Copper Age, the Bronze Age and then the Iron Age to fill the gap between the Neolithic and the beginning of the historical period. Actually, we cannot speak of an absolute chronology, since the different civilizations went through these various periods at different times and some went directly from the Neolithic period to a much more advanced stage, and even to modernity, because of contact with more advanced civilizations. However, it makes sense to use those terms in a relative sense to indicate the different stages that the various civilizations went through. Copper, as well as gold and silver, can be found in nature in metallic form (native copper). It is thus not necessary to obtain it from minerals containing copper oxide. It is for this reason that the oldest copper artefacts date back to 10,000 years ago, while traces of copper obtained from its oxides (malachite and azurite) are about 7,000 years old.

Copper is a rather ductile metal and it is therefore possible to work native copper by hammering it without having to smelt it. With the development of furnaces capable of reaching its melting temperature (1085°C), copper utensils such as axes and plowshares, as well as weapons, could be made. However, it must be said that most of the tools were still made of stone even in the Copper Age, and it took millennia for the use of this metal to spread.

Bronze is obtained by adding tin to copper. Bronze has a lower melting temperature (down to 800°C depending on the tin content), but it is even more difficult to work (or cannot be worked at all if the tin content is high). Bronze was used both to make tools and weapons and for artistic purposes; however, its introduction was quite slow even in the latter case.

Apart from bronze, brass (copper and zinc instead of tin) was also used in antiquity, but in much more recent times, i.e., after the end of prehistory. Furthermore, it is important to stress that there were problems with the supply of tin, particularly in the Mediterranean area, and this led to a reduction in the use of bronze.

The third metal to be used was iron, which was much more difficult to obtain than copper and its alloys. Pure iron has very poor mechanical characteristics and a melting temperature higher than that of copper (pure iron melts at 1,536°C). Actually, pure iron had no applications in the past and even today it is used only in electrical engineering. Its more regular use has always been in the form of the iron-carbon alloys; steel (with less than about two percent carbon) and cast iron (with more than two percent). Pure (native) iron is not found in nature, and thus iron and steel must be obtained by refining the minerals which contain it, mainly in the form of oxides. An increasing number of cast iron objects was found in sites dating back to the Middle Bronze Age and many believe that iron was a random by-product of copper refining. The difficulties in finding tin eventually led, at the end of the Bronze Age, to the replacement of bronze artefacts with iron artefacts. It should be noted that the iron of the time had lower mechanical characteristics than bronze and so this substitution was a makeshift solution almost everywhere. The systematic production of iron tools began in Anatolia, about 4000 years ago.

An important source of very precious iron was meteoric iron. Meteoric iron is always an alloy containing nickel and other metals, with compositions similar to that of modern stainless steel or alloy steel. It is thus possible to create tools, and above all weapons, of the highest quality from meteoric iron; much higher than the quality obtainable from other forms of steel available not only during prehistory, but also in historical times. The swords made famous by legends, such as Excalibur, would likely have been forged using the "stones from the sky" – even if the certainty that meteorites originate from space was not understood until the beginning of the nineteenth century.

However, we must explicitly state that metals have always been very valuable, not just in prehistoric times, and were used only in applications where the advantages justified the high cost, in particular for weapons. In the Age of Metals, tools were still mainly made of stone, and wood was by far the most widely used material for machinery and means of transportation. This scarcity of metals lasted until the Industrial Revolution.

References

1. Righetto G., *La scimmia aggiunta*, Paravia, Turin, Italy, 2000.
2. Harari Y.N., *Sapiens. From Animals into Gods: A Brief History of Humankind*, Harper, New York, 2015
3. Teilhard de Chardin P., *The Phenomenon of Man*, Harper. New York, 1959.
4. Morowitz H.J., *The emergence of everything*, Oxford University Press, Oxford, 2002
5. Diamond J., *Guns, Germs and Steel: The Fates of Human Societies*, W.W. Norton, New York, 1997.

2

From prehistory to history

2.1 TRANSMISSION OF INFORMATION

Conventionally, the transition from prehistory to history is linked to the introduction of writing. The period we refer to here as 'history' begins with the possibility of permanently documenting human events. This passage also has great significance from a technological point of view.

All living beings have the ability to record information; indeed, this ability is an essential feature of life. It is precisely the fact that the information necessary for the replication of the individual is coded in this way that allows living species to survive the death of the individual. Moreover, the possibility that mistakes (mutations) can be introduced in the replication process allows Darwinian evolution, based on such mistakes and natural selection. All living beings thus accumulate information within their own cellular structure and this function, in all the life forms inhabiting our planet, takes place in the very structure of DNA. In the case of eukaryotes – living beings whose cells have a nucleus – the DNA is located within the nucleus.

Owing to evolutionary advances, and the increasing complexity of the nervous system, living beings started to develop a memory capable of accumulating information, always internally, in a less precise but simpler way. However, while the information contained in the DNA is transmitted from generation to generation and is very stable (apart from mutations), the content of the memory of individuals is less stable, cannot be transmitted and disappears at the time of their death.

With the appearance of human intelligence, and above all of language, the possibility of exchanging information between individuals appeared. Information must always be stored in the memory of individuals, but the very fact that it could be communicated from one individual to another meant that it could be passed on from generation to generation. We have already mentioned the likelihood that the use of fire, which allowed humans to enjoy more free time from activities related to mere survival while increasing the time available for 'social' activities (essentially for the exchange of information between individuals), made this process easier.

In this way, civilizations differentiated, characterized by their specific cultural heritage, technologies and forms of art. Probably due to the needs arising from the new organizational complexity of agricultural civilizations as they developed with the Neolithic Revolution, important information needed to be recorded in a stable form on material capable of lasting for a long time. The same technology that had developed to produce ceramic vases and other implements and the bricks used for buildings meant that this problem could be resolved quite straightforwardly. Clay tablets could be engraved and then dried, or baked, using similar methods, to guarantee the required long-lasting properties. In other places, where buildings were made of stone instead of brick, such information could be recorded by engraving on stone slabs.

Other materials were then developed, derived from animal skins, plant stems or other by-products, on which to record information in a simpler and cheaper way using colored inks.

To be able to record information on any material support, a system of symbols suitable to represent the various concepts had to be introduced. Traditionally, it is believed that the recording of verbal languages in written form goes back to around 3200 BC (about 5,000 years ago) in Lower Mesopotamia, but isolated cases of pictograms date back to earlier periods.

It is customary to distinguish between logographic writing (commonly called ideographic, even if specialists prefer not to use this term), in which the symbols correspond, at least initially, to words or concepts; syllabic writing, in which the symbols correspond to syllables; and alphabetic writing, in which they correspond to the individual sounds. In the latter two cases, writing is therefore a recording of the language, and the meaning of written information can only be understood by those who know the language of the person who recorded it.

Humankind came out of prehistory with the introduction of writing, but it is worth pointing out that the transition from prehistory to history took place at different times among the various populations. Slowly, writing caused important changes in the very way in which the populations that had entered this new era lived and reasoned. In particular, processes such as abstraction, logic, analysis, classification, synthesis and the formalization of new theories were strengthened by writing.

Writing thus caused a further acceleration of technological development, in addition to that linked with the new needs of life in large communities, to the increasingly complex social and political organization, to the centralization of wealth in a limited number of hands and to the increasing number of people who could be diverted from the primary tasks of food production.

The new form of information storage decreased the importance of direct memorization of information, allowing people to think with greater detachment and objectivity about what they had learned through the mediation of a written text [1]. Writing led gradually to the birth of literature, poetry and of the personal interpretation of knowledge in all fields, making individualism and nationalism possible. While all literary works tend to become a collective elaboration in an oral culture, roles such as the individual author, poet, philosopher (and then scientist) are made possible thanks to writing.

In time, some of the individual authors could become reference characters, or *'maître a penser'*, able to exert their influence for long periods, even centuries or more. Reproducing the work of others as if it were a personal elaboration was a frequent, if frowned upon practice until, with the passing of time, it became the crime of plagiarism.

2.2 TECHNOLOGY AT THE BEGINNING OF HISTORY

History began in the valleys of the great rivers, the Nile, the Tigris and the Euphrates, the Indus, the Yellow River and the Blue River, where great agricultural civilizations had developed following the Neolithic Revolution. These civilizations were based on the domestication of plant and animal species, which differed mainly due to geographical and climatic factors. After millennia of artificial selection, the various species were now significantly different to the original wild species, having now become a product of human action.

But humans eventually changed the environment much more deeply. The cultivation of plants, which could allow a much greater number of people to live than could otherwise be fed in 'natural' conditions, required careful management of the river water upon which those civilizations depended. The construction of dams and canals, their maintenance, their defense and, in general, their management, was vital for their survival.

The situation was different from zone to zone. The natural regime of the great rivers varied according to geographic and climatic characteristics and therefore the interventions required varied from civilization to civilization. Moreover, this was a delicate dynamic equilibrium because the climatic factors could change over time. At the beginning of the Neolithic Revolution, humanity had just emerged from a very severe ice age and the temperature began increasing. Later, warmer and colder phases alternated with humid and drier phases. In order to survive, it was necessary to learn to dig canals and to build dams, but also to develop an organization that would make it possible to build, maintain and defend them, coordinating the work of the thousands of people employed in this collective effort.

Stone working had gone through a development lasting for millennia, but now metal tools began to be used together with the stone ones. Agriculture required new types of tools, such as scythes and sickles of various shapes, ploughshares, hoes, and others. In many areas, where big trees could be grown, wood became more and more important for producing tools of various types, which necessitated not only learning woodworking, but also building other metal, stone or wooden tools suited for this task.

Various types of weapons, often not very different from the working tools or derived from the latter, also had to be developed. Armies were among the organizations that had to be created and armies needed weapons, built in large quantities and with well-defined characteristics. As the military organizations began to differentiate from civilization to civilization, so too did their weapons. In some cases, they were technologies that today would be defined as 'dual use' – an axe or a knife can be used as a tool and as a weapon, for example – while others were strictly military technologies – a sword, for instance, is only a weapon.

Fire had both domestic uses and uses that today would be considered as industrial. In the Neolithic period, humans were able to light fire at will, both by rubbing techniques or by producing sparks, and they were now mastering it. Life in increasingly populated villages and the use of flammable materials in large quantities made the danger of fire an ever-increasing one, however, particularly because agricultural products, which had to be stored for long periods of time, could easily catch fire and be destroyed. Among the forms of organization that society had to create, there were groups of people trained to prevent fire from causing serious damage as a result of even a momentary loss of control. In many cases, these groups took the form of paramilitary corps.

The technologies which required the use of fire included both ceramics and metallurgy, with the latter in particular requiring quite high temperatures. Bellows and other tools had to be introduced to increase the draft through the fireplace or the furnace far beyond what was needed for simpler uses such as cooking food and heating rooms.

This brief review shows that, at the end of prehistory, civilization was not only totally dependent on technology to the point that an error in the management of the technological support could lead to disasters of incalculable proportions, but was also highly vulnerable to events such as climate changes with consequent droughts or floods, changes in the balance of power between neighboring states, or external invasions. Moreover, internal disturbances could also endanger the organization of the society. However, we should not over emphasize this alleged vulnerability, because the great agricultural civilizations which followed the late

Neolithic, and those of the following centuries, remained stable for extremely long periods. They were able to overcome many crises while maintaining an average standard of life far higher than had been common in the previous phases of human history, and certainly much higher than that of the populations which lived outside these 'privileged' areas in the same period.

2.3 SIMPLE MACHINES

Traditionally, the devices we call *simple machines*, known since ancient times, are those able to amplify the force that a man (or another agent) can apply to any object. Conventionally we distinguish six simple machines: the lever, the wheel and axle, the pulley, the inclined plane, the screw and the wedge (see Fig 2.1).

Figure 2.1. Six simple machines. [Image from https://upload.wikimedia.org/wikipedia/commons/2/20/Six_Mechanical_Powers.png.]

Some of these devices, such as the inclined plane, were not always perceived as 'machines', which is why they were often not included in the list in the past.

Today, we know that the laws of thermodynamics state that in any energy transformation, energy is conserved, but due to the energy dissipations which are present in any device a certain amount of energy is always degraded into a form from which it cannot be recovered (thermal energy), if not at the expense of further energy expenditure. No machine can therefore multiply the energy which is applied to it and, in contrast, the usable energy produced by any machine is always less than that supplied to it. The laws of thermodynamics are a recent scientific discovery and were completely unknown in the ancient world. Indeed, the very concept of energy, as well as those of work and power, was unheard of, so that we cannot expect that anybody in ancient times could forward considerations about the efficiency of any machine or about its ability to multiply a force. Even the concept of force was not clearly defined.

Today, we define a device that does not have its own energy source as 'passive', and one that uses some form of energy as 'active'. The machines of the ancient world were all passive and were operated directly by a human agent or by some domestic animal, at least until the energy from wind or from moving water from a higher to a lower level began to be exploited. These simple machines could thus increase human force, but the increase in force had to be paid for by a corresponding (or rather, greater, because of energy dissipations) decrease in displacement. Using a lever, a heavy object can be raised by applying a force less than its weight, but it is raised to a lesser height in this way.

Similarly, by multiplying the force, the speed of movement is reduced. Conversely, a simple machine can be used to increase the speed of a movement at the price of decreasing the force applied. In all likelihood, the lever principle was already being unintentionally used by humans in the Paleolithic, and is even used occasionally by some animals in a less conscious way. The difference lies mainly in the fact that while a monkey may use a branch as a lever to move an object, only a human, even the most primitive, can cut a branch from a tree, adapt it to be used as a lever and maybe store it for a similar future use. One application of the lever is the shaduf (Fig. 2.2), a simple machine used to lift water from canals or wells to cultivated fields, palm groves, vineyards and vegetable gardens. Its use certainly dates back to around 4500 years ago and was particularly widespread in Egypt.

Thanks to the lever, a counterweight, usually a large stone, balanced a much lighter bucket in such a way as to allow the bucket to be lifted to a much greater height than that reached by the counterweight.

In the Paleolithic, even the wedge had already found some application; after all, an axe is nothing but a wedge used with greater speed. The inclined plane was also certainly used, at least in the construction of buildings to raise the materials, as can be deduced from many drawings dating back to about 5,000 years ago.

2.3 Simple machines

Figure 2.2. The shaduf, a lever used as a lifting tool, mainly for raising water. [Image from Edwards, Amelia B. *A Thousand Miles up the Nile*, George Routledge and Sons, Limited: London, 1890. p 73, ://upload.wikimedia.org/wikipedia/commons/6/63/Shaduf2.jpeg.]

The other simple machines – the wheel and axle, the pulley and the screw – (Fig. 2.1), are certainly more recent and their use was documented in Roman times. However, given that the use of ropes has been documented since the Lower Paleolithic, it is likely that the former two are actually older, even if we consider that ropes can also be used without employing machines of any kind.

About 5,500 years ago, an innovation occurred that would deeply affect the future of humankind, namely the introduction of the wheel for vehicles. At that time, the potter's wheel had already been introduced for the production of axisymmetric pottery vessels, as can easily be deduced from the traces left on the vases produced in this way. The oldest evidence of a wheeled vehicle is a stylized sketch, used as a pictogram on a tablet found in the temple of Inanna, in Uruk, lower Mesopotamia (Fig. 2.3a). This document dates back to just after 3500 BC.

All the early vehicles had some common characteristics. The wheels were discs obtained by joining three wooden planks (Fig. 2.3b) and the animals were attached by means of a drawbar consisting of a single central pole. This uniformity of

26 From prehistory to history

wheel construction and of the drawing system, compared with the extreme variety in the structure of the actual vehicles, has led many historians to believe that the wheel was actually invented (or rather, it was probably developed from successive processes over a period of time) in a well-defined place and spread to the rest of the world from there. In the various civilizations, artisans would have adapted the structure of locally constructed sleds to the imported wheel.

Figure 2.3. (a) Pictograms representing a sleigh and a wagon. From a tablet from about 3500 BC found at Uruk, Mesopotamia. (b) Copper model of a war chariot with four onagers harnessed. It was found in a tomb at Tell Agrab in Mesopotamia (third millennium BC). [Image from G. Genta, *Motor Vehicle Dynamics*, World Scientific, Singapore, 1997.]

It is not known where the wheel was developed, but it can be surmised that this occurred in lower Mesopotamia where, as already noted, the wheel was certainly used around 3500 BC. According to the traces which have been found, the use of the wheel spread slowly. We find vehicles on wheels in Elam and in Assyria in 3000 BC; in central Asia and in the Indus valley towards 2500 BC; on the high Euphrates about 2250 BC; in southern Russia and Crete in 2000 BC; in Anatolia in 1800 BC; in Egypt and Palestine around 1600 BC; in Greece and in Georgia around 1500 BC; in China in 1300 BC; in northern Italy around 1000 BC and in northern Europe a few centuries later. From the primitive representations, it is impossible to understand whether the wheels were mounted free-wheeling on the axle or were fixed to a rotating axle. It is likely both solutions were used, as still happens today among peoples who have preserved their primitive technologies.

Towards the year 2000 BC, the need to build light carts, particularly in the case of war chariots, probably led to the development of spoked wheels, which for this application represents enormous progress.

In many cases, even in very ancient times, wheels had a metal rim, or at least were hardened at their periphery. Some disc wheels had a wooden rim, in one or more pieces. Sometimes, the periphery was reinforced with copper studs, to reduce wear or perhaps to hold a leather cover in place. Certainly, many Egyptian war chariots had leather-covered wheels. In some representations, even very old ones, it appears possible to identify a metal hoop, and traces of wheels with metal hoops, generally in segments welded together and then shrunk to fit, go back to the first millennium BC.

It was the domestication of animals that made it possible to have an effective traction system for vehicles. In Mesopotamia, onagers were used both for transportation wagons and for war chariots. Oxen were also used for the former application.

With the introduction of the spoked wheel, there is evidence of the use of horses for pulling war chariots. It is unknown where the domestication and the use of horses to draw carts took place, although the scant archaeological evidence available suggests that it occurred in north eastern Persia. From there, the horse spread throughout the ancient world, from China to Egypt and into Europe.

The limited knowledge of animal anatomy initially led to horses and onagers being attached to vehicles in the same way as oxen. Eventually, a pectoral was added to the yoke since, unlike oxen, equines cannot exert the pulling force directly with their shoulders. The breastplate pressed on the windpipe of the animal and, when considerable traction was required, tended to upset the animal, preventing it from exerting all the force it was capable of. Even in Roman times, the horse was unable to exert a pull of more than 600 N. The inability to understand and resolve this problem had a considerable influence on the development of ground transportation and of all the mechanisms moved by animals.

It should also be noted that iron shoes were not used for horses in the ancient world. On very hard or slippery ground, they temporarily attached metal, leather or straw tools to the hooves of horses, mules or camels. It is believed that adequate metal horseshoes were introduced into the Roman Empire from the East in the second century AD, but their use became widespread only in the eighth century.

There is evidence that both two- and four-wheeled wagons were used, and it is unknown whether one predates the other. It is likely that the need to maintain a low axle load, due to the lack of constructed roads and the narrow wheel width, meant that two-axle wagons were preferred for transportation. War chariots, which had to be light and maneuverable, were always built with a single axle. There is no evidence of the use of wagons with an articulated front axle in ancient times, and there were certainly no suspensions or steering mechanisms of any kind until at least the beginning of the Christian era. Two-axle wagons therefore had poor maneuverability and to force them to turn on soft ground must have been a very difficult job.

28 From prehistory to history

The need for maneuverability and lightness was pushed to the extreme in war chariots, which were always built with two wheels. By the third millennium BC, the war chariot was already widespread and had assumed a fundamental importance in the wars between the various states that thrived one after the other in Mesopotamia. Not surprisingly, the military power of a nation was measured by the number of chariots, as it is sometimes measured today by the number of nuclear warheads. Egypt and the states of Mesopotamia were rivals in the construction of lighter and faster chariots in increasing numbers. An Egyptian chariot of the fifteenth century BC is shown in Figure 2.4. It is undoubtedly the best that could be built in this field and, apart from improvements in details, remained unsurpassed for centuries. The war chariot lost its importance with the rise to prominence of cavalry.

Donkeys were probably used both as pack animals and for the transportation of people as far back as the third millennium BC, while in the second millennium, horses were occasionally used to transport people on their backs. The lack of technical implements necessary to give the rider a stable journey (saddle and stirrups) greatly limited the use of horses in this way for centuries. In the first millennium BC, cavalry gradually replaced war chariots throughout the Middle East, North Africa, and Europe.

Figure 2.4. Egyptian war chariot, from a tomb near Thebes (XV century BC). [Image from G. Genta, *Motor Vehicle Dynamics*, World Scientific, Singapore, 1997.]

After this period, chariots similar to those used by the armies of earlier times ended up being used only in sports competitions, which became extremely popular, at least in Greece and then in the Roman world. The use of wheeled vehicles for transportation was hampered by the lack of adequate roads. Sea or river transportation was also preferred wherever possible, particularly for heavy or bulky objects, because of a lack of understanding of how to use animals.

2.4 CRAFTSMEN AND SLAVES

As already stated, the two major advances which conventionally mark the end of the prehistoric age are the written recording of information and the development of simple machines. At the core of both innovations, however, was an even more radical and disruptive change: the beginning of the economic diversification of society.

The dissemination of writing postulates the existence of a specific social category, responsible for quantifying and recording the production processes carried out by other individuals. The same applies to the introduction of working tools. To replicate tools on a large scale, a clear division of tasks is needed between those who deal with primary goods, i.e. food, and those who make auxiliary objects such as pottery and other implements. Alongside the technology, in short, came the rise of the specialist technician, or craftsman.

At the same time, and as a result of the same process of social diversification, a new phenomenon emerged which was bound to have a huge impact on the technological world of antiquity: slavery.

Although there is no archaeological evidence that allows precise dating, it seems that in the Paleolithic and Mesolithic the practice of slavery was virtually unknown. In a society based on hunting and gathering, the ownership of other individuals would have involved a greater burden – even if only at the level of pure subsistence – compared to the benefit offered by their contribution to such tasks. With the advent of agriculture and breeding, things changed radically.

From a microeconomic point of view, these two new activities have a marginal cost that is much lower than the previous ones. Once they have been started, obtaining an additional unit of product costs much less in terms of labor than it did with hunting or gathering. It can therefore become more expedient to increase production volumes by employing more people. A farmer's working day is in fact much more profitable than that of a gatherer or a hunter. The so-called "break-even point" between marginal costs and marginal revenues[1], that is, the level of saturation beyond which it is no longer convenient to increase production, is very high. The point is not just limited by what is convenient and can often be stated in terms of bare necessity. For example, irrigating land can involve a number of civil engineering works which greatly exceed the possibilities for an individual, and which tend to be similar whether cultivating a small plot of land or a much larger area.

From this point of view, the diversification of work between farmers and artisans involves a significant problem, since it takes many people away from the

---

[1] The marginal cost is the additional cost required to increase the production by one unit; the marginal revenue is the gross increase in revenue caused by an increase of production by one unit.

countryside while leaving the needs of the community unchanged. The solution came from the frequent wars that followed the birth of the first city-states. By resorting to force, it was possible to obtain a large number of prisoners, who could be enslaved and employed in productive activities at merely the cost of their sustenance. In an economy based on agriculture, slavery became a cheaper alternative to killing the prisoners or using them as food. Cannibalism was a common practice in prehistoric societies characterized by the chronic deficiency of proteins of animal origin, as happened in New Zealand, where it was practiced even in fairly recent times [2].

Evidence of prehistoric cannibalism was also present in the Mediterranean area, as confirmed by the Homeric poems, where this practice is attributed to primitive, non-Greek civilizations which did not practice agriculture. In the Odyssey, for example, Ulysses first lands on the island of Cyclops Polyphemus – a shepherd who captures and devours the intruders – and later also in the land of the Lestrigons, a warrior people who destroy the Achaean fleet, kill the prisoners and feed on them [3].

Slavery had become widespread during the Neolithic, but from a formal point of view, we must wait until the second millennium BC for the effective institutionalization of the phenomenon. In this sense, one of the first documents in our possession is the Code of Hammurabi (eighteenth century BC), which identifies slaves with the term *wardum* and attributes to them a subordinate role to both the so-called 'semi-free' and 'fully free' men. It is clear that this is the recognition of a situation that has been widespread for a long time, and not only in Mesopotamia. Almost all ancient civilizations made extensive use of slavery.

It is worth pointing out that, in this context, 'slave' does not necessarily mean a particular legal form but applies more generally to the figure of the subsistence worker. In fact, there were many differences from country to country. In the Egypt of the pharaohs, for example, one could not speak of actual slavery, except in the case of prisoners of war.

Contrary to the image coming from Greek sources, almost all the people working on the great monuments of the Nile valley were free men from all points of view. They were paid and protected by actual 'trade-union rights'. Egyptian peasants, however, were obliged to work for several months a year on public works and were also subjected both to a taxation of one fifth of any products obtained during the rest of the year and to periodical arbitrary confiscations. They may have been free, but, as the Egyptologist Ricardo Caminos said, they were "*constantly between abject poverty and utter destitution*" [4].

The role of slavery was not limited to the agricultural sector alone. Technological evolution and the start of specialized craftsmanship soon made it necessary to resort to slaves in this context as well, for unskilled, dangerous or repetitive jobs. At first, slave labor was employed in public buildings, in particular in the construction and maintenance of irrigation works and buildings commissioned by the ruling class. Following the spread of the use of metals, slave labor was also used in mining operations for the extraction of raw materials. The division of tasks between technicians and slaves was immediately very marked. In construction, the most complex jobs were carried out by specialists, sometimes referred to as architects. These individuals operated independently, were well paid and responded directly to the highest authorities. In Egypt, for example, it was the pharaoh himself who oversaw certain technical decisions. Under the coordination of these technical managers, there was a complex hierarchy made up of masters, artisans, workers and, finally, slaves. The scenario was widely different in classical Greece, with the advent of an early form of private initiative alongside the planned state economy. With the decreasing importance of central power, citizens found themselves for the first time with resources that exceed the threshold of subsistence. It was a surplus which constituted the birth of commercial enterprises. Due to the need to address a chronic lack of raw materials, many expeditions crossed the Aegean Sea and the entire Mediterranean, sometimes even establishing new settlements independent from the motherland. Soon, a similar private initiative also involved the technology sector, with the birth of the artisan workshop.

In the fifth century BC, a significant portion of the population of Athens worked in this sector and specialization proliferated. In the comedies of Aristophanes, the manufacturers of scythes, those of helmets, and also of crests, spears, trumpets and armors are mentioned. These were self-employed professions, exclusively dedicated to the realization of a single product [5].

From an organizational point of view, the chief craftsman, in some cases the manager of an actual shop, dealt directly with the customer, often a member of the country's elite. People of lower rank worked under him. In classical Greece, almost all free men owned at least one slave, and craftsmen used them in their shops. A confirmation of this comes from Aristotle (384 BC – 322 BC), who wrote in his *Politica* about the possibility of constructing what we would now call 'automatic machines', or even robots: *"If every instrument could perform on command, or rather by itself, its function, as Daedalus' artifices moved by themselves, or as Hephaestus' tripods spontaneously performed their sacred work by themselves; if, for example, the spools of the weavers could weave by themselves, the master of art would no longer need help, nor would the master need slaves"* [6].

Thus, the artisans had helpers and routinely used slaves. However, there was a definite structural limit to growth, a limit that prevented the emergence of a real industry. Despite their widespread distribution and their more than significant importance in the Athenian economy, artisan shops never managed to obtain financial loans.

The city banks, which in the fifth century were quite widespread, were not used to investing their capital in this type of activity. In modern terms, this policy is difficult to understand. The average bank interest fluctuated between 10 and 24 percent, while the money invested in shops could yield up to 30 percent [7]. Why this choice?

One of the motivations is certainly of a sociological nature. In spite of their relative familiarity with the ruling class and of their economic wealth, artisans were considered by the intellectual class as individuals of lower status by their very nature, as they put their intelligence at the service of the transformation and use of matter. Again, in his *Politica*, Aristotle says that, at a universal level, *"he who can forecast with intelligence is a ruler by nature, a master by nature, while he who can toil with his body is slave by his nature"* [8].

Given the strong dualism between matter and spirit which pervaded ancient Greek culture and philosophy, the craftsman had a low social rank independent of his income. In classical Athens, 'applied' knowledge, or *techné*, was not even considered as a true form of knowledge and was not worthy of being compared either to mathematics and geometry, or *episteme*, nor to humanistic culture, or *sophia*.

It is significant that the term *demiurge*, which in Greek means craftsman, was used by Plato (428 BC – 348 BC) in his *Timaeus* to designate a sort of lesser divinity, who gave shape to primordial matter to form the Earthly world. It is to his work that we owe the serious imperfections we see around us: the demiurge tried to replicate the supernatural realm, but precisely because the craftsman's technology was a far from perfect science, his creation was clearly inferior to the original. The same term was then used with an even more derisive meaning by the Gnostic sects of the early Christian centuries, which attributed a tyrannical attitude towards humankind to the demiurge. In short, craftsmen could perform a useful or even a necessary function, but this was still a lower task, not worthy of being remembered.

This is why we seldom know the name of the craftsmen from this period responsible for creating the artefacts we admire in museums and the technologies that allowed the ancient world to develop materially. Perhaps the only exception was that of the architects, whose work was often of high ideological value and with little practical or economic fallout and was raised to the rank of art.

There is a great distance between art and craftsmanship. Art is not *techné*, so the architect's name is worthy of being handed down to posterity. This is the case of Phidias (about 490 BC – about 430 BC) in fifth century Athens and, two millennia earlier, of Imhotep (ca 2700 BC), who was a vizier and an architect during the reign of Pharaoh Djoser.

The son of an architect, Imhotep had a multifaceted personality, to the point of being called *"Chancellor of the Pharaoh of Egypt and second only to him, doctor, administrator of the Grand Palace, of hereditary nobility, high priest of Heliopolis, architect, chief carpenter, chief sculptor and chief potter"*. He was probably the

2.4 Craftsmen and slaves

creator of the first temple of Edfu and of the stepped pyramid of Saqqara (Fig. 2.5). In this case, however, he was an individual of very high noble lineage, active both politically and professionally. It is for this reason that his activity as a doctor, accurately documented, allowed him to ascend after death to the community of gods as god of medicine.

Figure 2.5. Bronze statuette of Imhotep, currently at the Louvre. [Image from https://upload.wikimedia.org/wikipedia/commons/9/95/Imhotep-Louvre.JPG.], and stepped pyramid of Saqqara. [Image from https://upload.wikimedia.org/wikipedia/commons/0/09/Saqqara_pyramid.jpg.]

Apart from these rare exceptions, in the ancient world the class of 'technicians' was considered much lower status than that of philosophers. We know quite a bit about Archimedes (about 287 BC – 212 BC) only because of his involvement in the wars against Rome, whereas with Heron, we do not even know exactly in which century he lived.

Aside from the existence of this social limit, the development of ancient technology was also strongly constrained by an intrinsic factor. The work of artisans, technicians and inventors of antiquity was of a completely empirical and non-scientific nature. The construction of an instrument never presupposed the knowledge of the physical laws that governed its function; on the contrary, the search for such universal rules was completely neglected.

Almost nobody wondered why an object worked in a certain way, inquiring at most only how it was possible to make it work. Even in the rare cases in which an answer to the first question was sought, this did not go beyond a series of explanations of a magical nature. In many cases, it was the craftsmen themselves who protected their technologies by keeping them secret and did their best to muddy the waters by encouraging the most imaginative explanations.

This was particularly true in the field of metallurgy, which has always been surrounded by an aura of mythology. Even today it is not known how certain particularly valuable artefacts, such as the swords of Damascus for example, were made. Certainly, the ironsmiths of the Iron Age, late antiquity and the Middle Ages were well aware of the heat treatments and processing methods, but they had no real explanation of how such methods worked that began with the properties of the material itself.

A typical example is the nitration of steel, a process that enriches the surface of objects made of steel with nitrogen, making them particularly hard. The nitrogen must be in atomic form and is obtained by the dissociation of ammonia or other organic compounds. To treat the swords, the red-hot metal was sprinkled with urine, to which a supernatural property was attributed. Alternatively, for the finest swords, many populations used to immerse the red-hot blade in the body of a particularly valiant prisoner, so that the virtues of the defeated enemy passed into the weapon; a mythical justification in full accord with the magical beliefs of the times.

It is obvious that these explanations did not lead to the fine tuning of the technological process, preventing above all the ability to replicate it accurately. The lack of instrumentation, mainly for measuring time and temperature, was particularly important and meant that the use of 'magic' was almost a necessity. The temperature was judged by the color of the incandescent metal and, in order to measure the time with any regularity, they had to rely on sequences of words pronounced with a particular rhythm. It was thought that if a process had worked by rhythmically pronouncing a certain formula, then it could be repeated by pronouncing the same words, which gradually became an indispensable magic formula.

In the absence of a real understanding of the causes, and with the constant desire for secrecy, progress was necessarily slow and strongly linked to discoveries made by chance. In the same way, the death of a particular craftsman could cause the complete loss of a technology, which would have to be reinvented from scratch.

Nevertheless, the great technological achievements obtained in this pre-scientific period were certainly not negligible and retain their charm even for contemporary people. Some achievements of ancient architecture, such as the pyramids of Giza, continue to inspire wonder to the point of feeding a growing number of legends about a supposedly lost esoteric wisdom of antiquity. In the absence of adequate scientific support, such knowledge tends to be imaginatively attributed to contacts with extraterrestrials, or to the legacy of very advanced pre-existing civilizations which disappeared without leaving any trace. In this case, as for the enchanted sword mentioned above, the reasoning is not based on any documented foundation. Irrationalism, moreover, is certainly not a phenomenon confined to antiquity.

2.5 THE INVENTION OF MONEY

"*Money is not something absolute: it's a technology that has changed over the millennia to satisfy a need*". According to economists Matthew Bishop and Michael Green, even the content of our wallets, while not having a mechanical nature, belongs to the great technologies conceived by mankind [9].

From a purely natural point of view, money is not a simple raw material. It is certainly not so today, when money is represented by simple pieces of paper and digital data, but it was not so even in ancient times. Actually, even gold and silver have a very limited intrinsic usefulness, to the point that the Aztecs – who had never been in contact with the Eurasian civilizations prior to the invasion – were amazed by the importance Spanish conquistadores gave to these metals.

The innovative and technological nature of money lies not in its physical structure, but in the abstract theoretical elaboration that underlies it. To attribute value to gold, subsequently to pieces of paper, and today to digital bits, is a surprising act of collective faith.

The historian Yuval Noah Harari, tracing back the history of this invention since the time of the early civilizations, observes that the first step towards the creation of money occurred at the time of the Sumerians. With the specialization of human activities and the diffusion of craftsmanship described in the previous section, the need for an easier and more immediate transaction than bartering became clear [10].

When performing an exchange between just two categories of goods – for example, wheat and meat – it is enough to apply the laws of supply and demand to determine how much wheat corresponds to a steak. With the spread of hundreds of different goods and services, several problems arise, not only due to the possibility of exchanging the various products, but also to the different needs of the parties involved.

As Harari noted: "*In a barter economy, if 100 different commodities are traded in the market, then buyers and sellers will have to know 4,950 different exchange rates. And if 1,000 different commodities are traded, buyers and sellers must juggle 499,500 different exchange rates! It gets worse. Even if you manage to calculate how many apples equal one pair of shoes, barter is not always possible. After all, a trade requires that each side wants what the other has to offer. What happens if the shoemaker doesn't like apples and, if at the moment in question, what he really wants is a divorce?*" [11]. Looking for a lawyer for such a triangular exchange would seem the obvious solution, but it would be necessary to find a lawyer who accepts apples as a compensation and does not want other goods or services, to avoid extending the transaction to infinity.

To overcome this problem, typical of complex economies, in Sumerian cities the use of barley grains as a common currency for commercial transactions was widespread. Barley was the basis of the daily diet and was a long-term asset. Contrary to metals, it had an actual, immediate value. However, those who possessed thousands of measures of barley did not buy it to eat it. Barley was accumulated by the rich not so much for its intended purpose, but for the fact that this commodity was universally accepted as a currency of exchange. The definitive point at which the currency of exchange was superseded was in the Babylonian age, with the transition from barley to silver, a material with little use in daily life.

With a level of abstraction unthinkable at the time of the hunter-gatherers, in this era we moved from the concrete and tangible concept of 'goods' to the abstract and potential concept of 'wealth'. As Felix Martin states[2], money is a "*social technology*", able to quantify such wealth and measuring an individual credit that can be transformed, on request, into any concrete good or service available on the market.

From the historical point of view, it was not a punctual invention that occurred at a precise moment and place. The use of this 'social technology' flourished in parallel in many regions of the world, due to independent elaboration, as a natural consequence of the increasing complexity of economic relations. Moreover, unlike mechanical technologies which are invented by one person and used by the individuals with whom he decides to share them, money is necessarily a collective elaboration.

It is the only true 'democratic invention' of human beings, elaborated and enjoyed by the whole community. To be precise, it would be more accurate to speak of it as a 'unanimity invention'. Unlike what happens in a democracy, it is not enough that money is accepted by the majority of the population. If even a significant minority of the human population stops accepting that a unit of money (whether it be a bowl of barley, a gold or silver coin, a sheet of paper or a digital bit) corresponds to a credit right that can be transformed into any good or service chosen by the owner, these symbols would instantly lose their value. Why accumulate money if it is no longer certain that this allows my daily subsistence? Even the risk that just ten percent of sellers may not accept the exchange would open huge cracks and chasms in the confidence in currency, because we would try to accumulate, in parallel to money, another object that would be acceptable as a commodity exchange to this minority, extending the demand for this commodity and so transforming it into a competing currency. Subsequently, an exchange rate would be set between the two currencies, thus restoring the universal acceptance of money in one of the two forms.

Evidently, like any form of plebiscitary consensus, trust in money must not only be widely spread and accepted by the community, but also, and above all, imposed and granted by the power of a central political authority. The use of silver

---

[2] It is one of the basic assumptions contained in his book *Money: The Unauthorized Biography- -From Coinage to Cryptocurrencies*, Bodley Head, London 2013.

as compensation for certain crimes, for example, was imposed by the code of laws stated by the Babylonian King Hammurabi.

What made this link between political authority and the circulation of money inseparable was the introduction of the first silver coin by the king of Lydia Aliatte in the seventh century BC. Previously, silver and gold were freely measured by weight at the time of trade. The coin, on the other hand, represented a standard unit of measure guaranteed by the state. It bore the effigy of the monarch to guarantee that every coin actually reflected the same value indicated on it without the need for further measurements. Given the presence of a royal signature, falsifying the contents of a coin was thus a political crime of *lèse majesté*, and a challenge to the authority punishable with death. Coins are not only a practical and safe way of circulating money, but they are also the only officially recognized currency. By introducing coinage, the state outlaws any other form of commercial transactions, acquiring the monopoly on the issue of money.

Money changes its function. At an individual level, it continues to represent individual wealth, but at the same time from a macroeconomic point of view, it is a fraction of the overall wealth of the state that issued it. In other words, the value of the total quantity of money in circulation coincides with the economic strength of the central authority. With the fluctuation of the relations between these two quantities, it becomes possible to speak of inflation and deflation.

In short, money is no longer an exchange asset like any other, but becomes a technological device representing a unit linked to the state economy[3]. This was an evolutionary path that led to the birth of the first complex economic system of the ancient world, that of the Roman Empire.

Above all, in addition to already representing a new revolutionary technology in itself, money would soon acquire a fundamental role in the development of all other human innovations. The link between technology and economy, as we will see later, is the basis of the contemporary Western world.

References

1. Goody J., *The Domestication of the Savage Mind*, Cambridge University Press, Cambridge 1977.
2. Diamond J., *Guns, Germs and Steel: The Fates of Human Societies*, W.W. Norton, New York, 1997.
3. Homer, *Odyssey*, book IX (Poliphemus) and X (Lestrigons).
4. Wilkinson T., *The rise and fall of Ancient Egypt*, Random House, New York 2010.
5. Pekàry T., *Die Wirtschaft der griechisch-römischen Antike*, Steiner, Wiesbaden 1979.
6. Aristotle, *Politica*, I, 4, 1253b 33 – 1254a 1.
7. Reference 5, p. 45.

---

[3] With the passage of millennia, it became possible to replace the coins made of precious material with notes made of paper, whose value remained guaranteed by the state.

8. Reference 6, I, 2.
9. Bishop M., Green M., *Face it: Money is Technology, and we can do better than gold*, in "Business Insider", online, 2013. See also Green M., *In gold we trust? The future of money in an age of uncertainty*, e-book published by The Economist Newspaper Limited, 2013.
10. Harari Y. N., *Sapiens. From Animals into Gods: A Brief History of Humankind*, Harper, New York 2015, "The Scent of Money".
11. *Ibid*

3

Greek rationality

3.1 A UNIQUE PHENOMENON

In the previous chapters, reference was made to the birth of agriculture, the specialization of the work of craftsmen, technological development and the use of slave labor. All of this holds true for the Fertile Crescent, as well as for the Indian city-states of Harappa and Mohenjo Daro, the ancient Chinese dynasties Shang, Zhou, Qin and Han, the peoples of pre-Columbian America and many other less known cultures.

In all this, there is nothing capable of justifying – even if only in an embryonic form – the subsequent technological supremacy of the West compared with the rest of the world. When and where did this gap, which was bound to take macroscopic proportions, start?

It would be wrong to try to locate a turning point by considering only the history of material technology. China, for example, was the birthplace of a large number of impressive innovations throughout its history; not only paper and gunpowder, but also steel working, which reached an impressive level in the tenth century AD, many centuries before the Industrial Revolution began in Britain. In 1018, Chinese foundries produced about 35,000 tons of metal tools a year, and sixty years later the level rose again, reaching 100,000 tons [1].

The widespread nature of technological advancement is even more clear when the development of architecture is considered. From the Great Wall to the pyramids, there are many impressive monumental structures in many regions of the planet, which prove without any doubt that complex instruments and advanced design ability developed everywhere.

Although at first it might seem of little relevance, the peculiarity of Western civilization is to be found in the history of philosophy, and more precisely in the development of a rational concept of reality. The unique characteristic, which set Western civilization in a category of its own, was actually to go beyond the magical-irrational vision of the world and approach reality with a scientific attitude based on the principle of cause and effect.

The Greeks were the first to question the causes of natural phenomena systematically, *"teaching themselves to think"* [2]. In the 1970s, the American psychologist Julian Jaynes argued that even self-consciousness originated at that time [3]. The neurological structure of Homeric heroes, as well as that of the non-Greek Middle Eastern populations, would still be unable to conceive the ego, and therefore to think in a truly rational way, without the need to relate the self with a third subject, the divinity. It is a thesis that is hardly acceptable from a scientific point of view, but which was forwarded to explain the macroscopic gap between classical Greek rationality and the irrationalism of other civilizations. The case of the periodic flooding of the Nile is emblematic of this. Despite the development achieved by the ancient Egyptian civilization, it was the Greek historian Herodotus who formulated the first three hypotheses that tried to find a rational explanation for an event on which the inhabitants of the Nile valley had based their well-being for centuries.

The three hypotheses, built upon what had been forwarded by previous Greek visitors, were based on natural elements and on the physical-meteorological dynamics. Prior to this, the Egyptian priests and nobles had limited themselves to attributing the phenomenon to the generosity of the goddess Isis, without attempting to formulate any naturalistic proposal to explain it.

Greek philosophy is based precisely on this type of empirical thinking, even though at the beginning it had no practical and operational applications. As already stated, philosophy and the technology of craftsmen were, after all, at the opposite poles of a cultural dualism that pervaded – and limited – the entire Greek civilization. However, even if we cannot yet label this as applied science, it was in this phase that foundations were laid for its subsequent development. Thales of Miletus, for example (about 630 BC – about 547 BC), who is traditionally considered to be the first philosopher, tried to investigate the causes of earthquakes, the laws of geometry and the origin of the world.

The implications of this were huge. To state that these phenomena were due to material causes meant overcoming magical irrationalism and asserting that the entire universe instead followed well defined laws, which, starting from certain premises, always had repeatable effects. It is no coincidence that ancient Greeks described reality using the term 'cosmos'. The word *kósmos* means 'order', or rational equilibrium. These early philosophers maintained that everything that surrounds us was rationally organized and could therefore be explained rationally. From a physical point of view, everything could be traced back to one or more natural principles. Compounds were then generated by combinations of these simple elements, becoming increasingly complex up to the level of humans and their daily reality.

Depending on the singular or plural nature of these basic elements, it is usual to distinguish between monistic philosophers such as Thales, Anaximander (about 610 BC – 545 BC) and Anaximenes (about 586 BC – 528 BC), and pluralists like Empedocles of Agrigento (495 BC – 444 BC), to whom we owe the most complete formulation of this model. The most interesting contribution comes from another pluralist, Anaxagoras (about 496 BC – 428 BC). Like Empedocles, he too described reality starting from the aggregation of innumerable primary seeds; the atomic elements.

Subsequently, however, he also considered the *nous*, the rational order that organizes these elements, coming to attribute a divine nature to this cosmic equilibrium: "*It is infinite and absolute master, and to nothing it is mixed but only he is in himself. (…) It is the purest and lightest of all things: it has total knowledge of everything, and the greatest power over everything*" **[4]**.

Such reasoning brought the rational arguments of the philosophers into conflict with the religious world for the first time. Anaxagoras was charged with impiety by an Athenian court, and was condemned to death for advocating a kind of monotheism, in contrast to the existence of the Olympus gods. Pericles stepped in and succeeded in obtaining his release, on condition that Anaxagoras left the city of Athens in exile forever. Years later, Socrates (470 BC – 399 BC), was less fortunate, as is well known.

Interestingly, the theme of this conflict had nothing to do with the secular nature of science. The idea that everything is rationally explicable and obeys universally valid laws did not conflict in any way with the existence of one or more divine principles, but only with the archaic view of a magical and mysterious universe, governed solely by the arbitrary will of the gods. The turning point would only come – as will be described later – when the Greek rationality met a religion based on the existence of a cosmic order; the Judeo-Christian order.

3.2 HUMANS, GODS AND TECHNOLOGY

Technology soon came under the scrutiny of the new rational thinking of Greek philosophy. Where did technology and scientific innovations come from? According to a verse attributed to the poet Epicharmus (about 524 BC – about 435 BC), there was no doubt: "*It is not that humans found technology. It is the god who taught them it*" **[5]**.

Actually, Olympic mythology, reworked by Plato in his *Protagoras*, conceived an evolution articulated on three levels **[6]**. At each stage, there was the intervention of a divine, or at least superhuman, figure. First, after the creation of the mortal species, the gods entrusted the titan Epimetheus with the task of distributing among them strength, speed, stamina and other physical gifts. By mistake, the titan distributed almost everything to animals without any reason, leaving humans unable to defend themselves. In order to find a remedy to this error, which was by then irreversible, Prometheus, the wise brother of Epimetheus, had to intervene at a later time. The titan stole fire from the gods and gave it to humankind. Together with fire, Prometheus also took the technical knowledge *'techné'* from Olympus. In short, humankind received the ability to transform the hostile natural environment into an artificial one, rebalancing the relationship with the animal world and escaping extinction. However, a socio-economic problem remained which prevented humans from exploiting these qualities. Technology, as mentioned before, can reach its full development only thanks to the rational division of labor. After a violent phase of civil wars, Zeus sent Hermes to establish *"respect and justice"*, as required to reach a political balance.

According to Greek religion, every achievement of human reason was bestowed by a superhuman being such as Epimetheus, Prometheus and Hermes. In short, rationality was only a small sphere, which the gods created around every human to enable them to defend themselves against the physically stronger and faster animals. However, the general rule that governed the universe was not rationality, but the chaotic inconstancy of the gods, who did not comply with any general rule and manipulated the lives of humans as they pleased. In this sense, the Homeric myths were emblematic: human enterprise succeeds or fails only to the extent that the quarrelsome and manipulative gods arbitrarily decide. When the same action is repeated in the same conditions, the results may be opposite.

This concept was contradictory to that of the philosophers, who instead claimed that it was possible for the human mind to investigate the entire universe. While for the poet Epicharmus technology came from the gods, Xenophanes of Colophon (570 BC – 475 BC) observed *"the gods didn't reveal everything to mortals since the beginning, but over time, they found the best things [by] looking for them"* [7]. It was humans who invented the technologies, which had nothing to do with the divine world. Xenophanes went as far as to challenge the anthropomorphic nature of the gods, radically rejecting the concept of Hephaestus as a blacksmith and Athena as lawmaker.

By negating these human traits of the gods, it finally became possible to investigate discoveries and innovations. Xenophanes published various books on the subject, and we know for certain, for example, that he attributed the invention of money to the Lydians. For the first time, the technical achievements of humankind were considered worthy of being remembered.

Theatre was also influenced by this new point of view. In his *Antigone*, Sophocles (496 BC – 406 BC) celebrated human intellect by stating that *"Many wonders are in the world, but no wonder is equal to man (...). With his arts, he tames the wild beasts that live in the mountains and bends under the yoke the horse with its thick mane and the strong mountain bull. He learned how to speak and to think as fast as the wind and to take civil commitments, he learned to take shelter from the bites of frost and stinging rains. Full of resources, he is never taken unprepared by what awaits him, he has found a remedy for irremediable ills. Only to death he cannot escape. Absolute master of the subtle secrets of technology, he can do evil as well as good*" **[8]**.

In short, technology was something typically human. Even the Athenian philosopher Anaxagoras aligned himself with this concept prior to being condemned to exile for impiety, reworking the myth of Epimetheus and Prometheus in a rationalist way. Having understood how, before the birth of civilizations, man was effectively helpless when faced with other animals, Anaxagoras argued that what made the difference was not the divine gift of fire, but the four natural qualities of experience, memory, knowledge and technology. What caused the development of intellect and technology was therefore the biological structure of the human being. According to Anaxagoras, humans had at their disposal a precision instrument with which no other animal was provided: their hands. For the first time, in stating the importance of this organ as a dividing line between humanity and the animal world, philosophy and technology found a common ground after centuries of opposition.

Knowledge (*sophia*), of which Anaxagoras spoke here, was no longer abstract thought, but the practical ability to use the hands to modify the surrounding world. Eulogizing the craftsman's technique to such a magnitude, however, did not fit well with the ancient society and with the strong aristocratic prejudice towards manual work. As the historian of philosophy Giuseppe Cambiano observes, "*Anaxagoras' sentence to exile from Athens has thus a new motivation, as he had indirectly supplied, with his biological-anthropological theories, an ideological support to the classes which were claiming democracy*" **[9]**.

Anaxagoras, however, did not remain an isolated voice. Democritus (460 BC – 370 BC), the inventor of atomistic physics, argued that in ancient times, finding themselves at the mercy of the stronger and faster animals, humans created a first system based on fear. Threatened by other creatures and dismayed by the meteorological and astronomical phenomena, they attributed the cause of all that could not be explained to the gods. Only after the development of society and language did humans succeed in developing a second model, this time based on experience. According to Democritus, technology was born in this second phase, the origin of which could be tracked back not so much to the gods as to the rational imitation of nature itself. "*We were pupils of the animals in the most important techniques: the spider in weaving and mending, the swallow in the construction of houses, the singing birds, the swan and the nightingale, in singing*" **[10]**.

However, not all the Greek philosophical world agreed with this new rational perspective. The charismatic Parmenides (515 BC – 540 BC), founder and master of the school of Elea, who stated that his philosophy was started by his encounter with Dike, the goddess of justice, reaffirmed strongly that the great truths came from the divine world, with humans having merely the faculty for developing them. In adopting a clear and radical perspective, Parmenides dismissed any attempt to give life to applied science. The object of philosophical investigation had to be the reality that lay beyond appearance, the Being. This divine entity was characterized by qualities unrelated to the world surrounding humankind: invariability, uniqueness, eternity, immortality and indivisibility.

The world in which people lived, constantly subject to transformations and changes, was instead the realm of appearance. Trying to study it and to understand its working, or even to transform it on a human scale, was a waste of time, and the thoughts of the wise had to be addressed exclusively to the Being transcending the world, which – as Being – was the only existing reality.

This thinking by Parmenides had a significant following and deeply influenced the ancient world through the studies carried out by the disciples of his school of Elea. After several decades, even Plato based his thinking on the assumption that all that surrounded humans was nothing but the corrupted copy of a distant original, and that the only truths worthy of being studied belonged to another dimension.

In short, study by philosophers had to be directed to things which did not change. Based on this assumption, a disciple of Parmenides in the fifth century, Melissus of Samos (about 470 BC – about 430 BC), severely attacked every attempt to promote the application of science to the reality of humankind. To do so, Melissus addressed his argument against the field of medicine, which had most benefited from the contributions provided by Greek rationalism.

3.3 APPLIED RATIONALITY: THE ORIGINS OF MEDICINE

In his *Stories*, the Greek historian Herodotus (484 BC – 430 BC) reported that the inhabitants of Mesopotamia "*having no doctors, bring their sick to the public square, and those approaching the sick express an opinion on his illness, if by chance they had the same symptoms or if they knew of someone who had them*" [11].

Things were very different in the Nile valley, where "*the country is full of doctors (…) of the eyes, the head, the teeth, the abdominal region and the diseases of uncertain localization*" [12]. Not surprisingly, the Egyptian doctors were particularly revered throughout the Middle East, to the point that the Persian emperor Darius used to rely on their treatments [13].

Moreover, the collection of traditional knowledge and magical superstitions typical of these populations had very little in common with the rational techniques adopted in the Greek world. The same Herodotus observed with shock that the Egyptians "*take care of their health with emetics and baths, because they believe that all diseases derive from the food with which we feed*" [14]. Then, if something could not be included in this basic model, it had to be traced back to the influence of the gods or other supernatural entities.

For the Greeks, in contrast, medicine immediately pursued the goal of systematically describing all pathologies by referring only to the natural world and using its own rational method. A clear declaration of such intent was expressed in the writing *The sacred disease*, attributed to Hippocrates (about 460 BC – about 370 BC), dedicated to the treatment of epilepsy that had traditionally been associated with the influx of the gods. According to the author, a serious doctor could not accept the existence of "*divine diseases*". If he did, he would be obliged to find a way to justify his own inability to be rational and prescribe improvised remedies. If he were successful, it would be attributed to the doctor-magician, whereas any failure would be deemed the responsibility of the gods.

"*(This disease) is by no means more divine and more sacred than the other diseases, but has a natural structure and rational causes: humans, however, held that, in some way, it had a divine origin for their inexperience and amazement, since in no way it looks like the others,*" the author stated [15]. In short, the civilization passed from the divine phase, dictated by fear, to a rational phase due to knowledge. The resonance with the thoughts of Democritus described in the previous section are evident.

The abstract rational method of the philosophers, however, found a practical application here, starting from the direct observation of reality: "*Opening the head you will find the brain moist and full of hydropic and ill smelling liquid, and then you will clearly understand that it is not the god who afflicts the body, but the disease*" [16].

Philosophical rationality aligned with the working technique of craftsmen, enriching and perfecting it. For the first time, in the case of medicine, it was possible to identify an applied science.

This assimilation to the technical disciplines, moreover, was proposed by Aristotle – who attributed the status of *demiurge*, i.e. craftsman, to the doctor – as well as by the followers of Hippocrates, authors of a methodological treatise on their way of operating, entitled *Techniques*. As it was applied to everyday life, Greek medicine did not claim universal validity, as happened with logics and philosophy. In the text *Techniques*, for example, the followers of Hippocrates observed how their diagnoses were effective only when they were exercised on a patient who was not completely overwhelmed by their disease [17]. As Giuseppe Cambiano observes, according to the ancient Greeks "*medicine has an autonomous field within which it can exercise its dynamis (efficacy): observations, experiments and analogies are the instruments that allow [them] to formulate a diagnosis. But, extrapolated from this field, medicine loses all possibility and effectiveness*" [18].

Moreover, in their methodological treatises, the Hippocratic doctors also pointed out that "*the craftsman's techniques, that depend on fire, are useless when the latter is absent, and are usable when it burns (...) the action ceases when an instrument is missing*" [19].

As mentioned in the previous section, with Melissus of Samos, the Eleatic philosophers in the fifth century challenged the reliability of medicine as a concrete discipline aimed at the earthly world, which in their view did not really exist. For example, based on the assumption of the unity of the Being, Melissus stated that it was impossible to break down a human being into a multitude of elements, as the Hippocratic doctors had done, because these parts would vanish in a single human entity. This criticism, which was destined to have a significant impact on the credibility of medicine in ancient Greece, was rejected by adopting an empirical position. According to the author of the treatise *Ancient medicine*, it was incorrect to start from a general hypothesis as the philosophers did and it was instead necessary to refer to the direct analysis of reality. Proudly claiming the status of technology for medicine, as a discipline that produced a benefit and a practical utility for humankind, the followers of Hippocrates denied equivalent status to those who looked to make *"a technology from the art of negatively criticizing other technologies"* [20]. This quite open attack was directed at the philosophers who were hindering the development of applied science.

However, there was a positive side to the influence of philosophy through the construction of a medical *odós*. This was a logical-rational method that espoused the progressive enrichment of the knowledge of the sector based on its previous achievements [21]. The journey towards scientific perfection was not a smooth one without stumbling blocks, nor was it free of moments of drastic relapse into irrationalism.

The most well-known case is certainly that of the double plague epidemic, which struck Athens in 430 BC and 427/426 BC. The disease hit a community already severely afflicted by the war against Sparta and left the scientific world of the time helpless. The historian Thucydides observed that *"not even the doctors, who did not know the nature of the disease, and treated it for the first time, were enough to cope with it. Indeed, they themselves died more than the others, since they approached the sick more than the others, and no human technique was enough against the plague"* [22].

The whole structure of scientific knowledge of the time was questioned: *"There was no particular medicine that could be used to heal: what was useful to one, to another was harmful"* [23]. It was an epochal phenomenon. According to Pliny the Elder, Hippocrates himself was among the doctors who rushed to the city in vain to try to defeat the disease [24].

Today, we think that it was probably a typhus epidemic. The episode marked a crisis in Greek rationalism and a temporary return to superstition. Having lost hope in a scientific solution to the problem, in 420 BC the inhabitants of Athens introduced the cult of the god Asclepius. Within a decade, so-called 'magical medicine' had therefore proliferated once more in the capital of the classical world.

3.4 GREEK NATURAL PHILOSOPHY

Greek philosophy derived from attempts to explain the world around them rationally. Its reflection on nature (in Greek, *physis*, from which we get the word physics) was established as a very important aspect of this from the very beginning.

3.4 Greek natural philosophy

The natural philosophers of classical Greece had achieved extremely important results in some sciences, such as mathematics and what we now call astronomy. At the time of Plato, for example, it had been established as fact that the Earth was spherical, a discovery generally attributed to Pythagoras or Parmenides, who lived in the fifth century. No other ancient civilization had managed to reach this result, which was counter-intuitive to the thinking of that age. Parmenides is also credited with the discovery that the phases of the Moon are due to its illumination by the Sun and the fact that the Moon is also a sphere. While the idea that the Earth was spherical had been consolidated since the beginning of Greek philosophy, the fact that the rotation of the celestial realm was only a perception and that what actually rotated was the Earth remained a controversial subject for a long time. To Aristotle in particular, the Earth (spherical) was motionless and the daily motion of rotation was imparted to the sphere of the fixed stars.

The military enterprise of Alexander the Great opened the Greek world to the East and the kingdoms, born from the shattering of the Macedonian empire, further favored this expansion. Thus began what is commonly called the Hellenistic period, to distinguish it from the previous period, characterized by the dominance of classical Greece. In particular, Ptolemaic Egypt, with its new capital Alexandria, became a cultural center of the highest importance. The first Ptolemies, and particularly Ptolemy II Philadelphus, attracted philosophers, writers, doctors and men of culture from all over the world to Alexandria, and the cultural policy of the Ptolemies was extremely aggressive. Understanding that knowledge was power, they decreed that any book entering the kingdom of Egypt should be handed over to the library to be copied and enter its bibliographical patrimony, with a copy to be given back to the original owner later.

With its lighthouse, library and other cultural institutions, Alexandria became the largest cultural center in the world, though it was not the only one. According to Pliny the Elder, for example, when Ptolemy V Epiphanes forbade the export of papyrus in the second century BC to strengthen the cultural monopoly of his kingdom, the king of Pergamon, Eumenes II, decided to produce a writing material from the skins of sheep in order to prevent the decline of the library of his city, thus introducing parchment.

The natural philosophy of classical Greece had little to do with what we now call science, particularly from a methodological point of view. The philosophers tried to assign qualitative explanations of reality in its entirety, trying to identify the deepest causes of the being, the unifying principles and the possibility of going beyond the visible effects. They studied the possible existence of meta-empirical causes, without worrying about defining the area in which the explanation was valid and, above all, without trying to validate the results obtained experimentally. However, during the Hellenistic period there were philosophers who introduced something closer to science as we understand it.

These scientists delimited their research field and formulated hypotheses that did not claim to be the truth, but provided a basis on which to build a theory that, while not purporting to be an explanation of reality, at least provided a model on which a particular aspect of it could be studied.

For example, when he introduced the heliocentric model of the solar system, Aristarchus did not think of presenting it as an explanation of the motion of the planets, but rather as a conceptual model to "save the phenomena", the common phrase of the time. It was a model that did not contradict the facts observed in the real world.

It was only with Seleucus of Babylonia that heliocentrism became more than a mere hypothesis, proven by the discovery of something approaching a true law of gravity [25]. He proposed that Earth was not the only object to attract heavy bodies towards its center (while rejecting light bodies upwards) as Aristotle claimed. Seleucus understood that all bodies were attracted to each other and that if the Earth attracted the Moon, which was in equilibrium between the force of attraction due to the Earth and the centrifugal force due to its motion (the comparison with the sling stone, in balance between the force exerted by the sling and the centrifugal force due to its motion, was made by Hipparchus), then the Moon also attracted the Earth. The Moon and the Earth both rotated, with a period of one month, around the center of gravity of the Earth-Moon system and it was this rotation of the Earth, together with the attraction of the Moon and of the Sun, which explained the tides (this "saved the phenomenon" of the tides). The proof of Seleucus' theory of the heliocentrism would therefore be linked to the observation of the tides, although we cannot be 100 percent certain of this since his books have been lost.

In the Hellenistic period it was therefore established that the Earth had three motions, each with its characteristic period:

- the rotation around its axis, with a period of one day,
- the rotation around the center of mass of the Earth-Moon system, with a period of one month, and finally
- the rotation around the Sun, or better around the center of mass of the solar system, with a period of one year.

An approximate chronology of the Greek and Hellenistic philosophers can be found in the Appendix, in Figure A.1. The exact date of birth, or death, of many of them remains unknown.

Astronomy was far from the only field that Hellenistic scientists devoted themselves to. Euclid studied geometry in a scientific way, basing it on five propositions to define his theories. His geometry concerned figures that could be drawn with a ruler and compass, notably ideal figures obtainable using ideal rulers and ideal compasses.

From Euclid, there was a proliferation of scientific disciplines, such as optics concerning the theory of vision, as well as mirrors and lenses, mechanics, pneumatics, hydrostatics, geography and, of course, astronomy. All these sciences were generally grouped under what Greeks called mathematics, which could be translated with exact science, as opposed to physics, which was natural philosophy.

Medicine also developed in many forms, from anatomy which was practiced by dissecting corpses, to surgery through cataract operations that became routine; and from psychiatry, with the scientific study of the interpretation of dreams, to physiology.

In 145 BC, Ptolemy VI's *Euergeies Physkon* drove the Greek philosophers and scientists away from Alexandria. Politically, it was an attempt to return control to native Egyptians and away from the Greek component of the Alexandrian society, but Hellenistic science received a blow from which it could no longer recover, even though many of the scientists working in Alexandria took refuge in other cultural centers of the Hellenistic world where they continued with their studies.

3.5 HELLENISTIC SCIENTIFIC TECHNOLOGY

Apart from astronomy, which also had important applications for navigation and cartography, the disciplines cultivated by the scientists of the Hellenistic period had strong technological relevance and, in turn, depended on technology[1]. In fact, this period was responsible for significant developments in experimental activities and consequently was characterized by the construction of measuring instruments which had never previously existed. In particular, the measurement of time saw important improvements. Archimedes and then Ctesibius worked to make the water clock more precise, and Ctesibius in particular introduced a constant-level vessel to make the water flow consistently over time.

A device found at Antikythera, a kind of mechanical calculator to compute the phases of the Moon and the motion of the Moon and the Sun, not only demonstrates the progress achieved in metallurgy and in the construction of gear wheels, but above all the understanding of how both ordinary and planetary gears work. The device, which includes about thirty gear wheels, actually includes a differential gear. The fact that the gear wheels and other parts of the mechanism were made of bronze clearly shows that, at that time, the use of metals for the construction of some parts of machines was spreading.

The invention of hydraulic pumps is also attributed to Ctesibius (about 285 BC – about 222 BC). These devices often included mushroom valves and their cylinders and pistons were quite precisely machined, showing the level reached by mechanics in the Hellenistic period.

The use of hydraulic pumps spread quickly. For example, the increasing size of the cities made the firefighters corps more and more important, while greater use of metals led to an increase in mining activity and therefore to the need to drain the mines. The first water wheels, which began to spread across many types of

---

[1] For further information about this topic, see L. Russo, *The Forgotten Revolution*, cit.

50 Greek rationality

industries in the Hellenistic period, were initially used to pump water from the mines. They were then used to drive the flour mills and the bellows in foundries, or for the hammers used in fulling (thickening) wool or metalworking.

Ctesibius used his knowledge of hydraulics and pneumatics to create a hydraulic organ, the first musical instrument to use a keyboard. In the hydraulic organ, the pressure in a tank was kept constant by the water level and was used to push air through pipes, where reeds produced the sound. Over the following centuries, the hydraulic organ spread throughout the Roman world, producing the 'soundtrack' that accompanied the shows and the public life of the empire.

Archimedes introduced the screw and then the screw press which, among other things, permitted developments in the food industry, especially for the production of wine and olive oil. Archimedes also introduced the cochlea, or Archimedes' screw, (Fig. 3.1), a device to raise water that guarantees a low head but a substantial flow, or to move loose materials such as sand. It still has many industrial applications today.

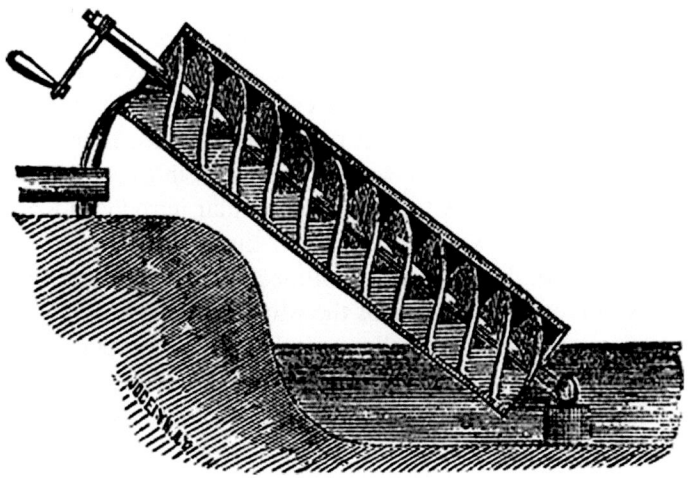

Figure 3.1. Cochlea, or Archimedes' screw, attributed to the latter. [Image from https://upload.wikimedia.org/wikipedia/commons/8/82/Archimedes_screw.JPG.]

We know quite a lot about Archimedes, mainly for his involvement in the wars against Rome, but his works have been almost completely lost. We do know that he made fundamental contributions to mathematics and theoretical mechanics, yet while he is known as a brilliant inventor, he has often been perceived as a strange and eccentric character. Perhaps the most famous example is the story (reported in *De Architectura* by Vitruvius) that he ran from his house naked screaming

"Eureka!" (I found it) after discovering the so-called Archimedes' principle more or less by chance while taking a bath. Or rather, after having understood how to solve the problem that had been posed by Hieron concerning the true composition of a golden crown that the tyrant suspected had been made from a less precious metal. In fact, Archimedes had developed a mathematical theory concerning hydrostatic force and understood what density was and how it was possible to measure it, without having to experience the movement of the liquid due to the immersion of a body while taking a bath.

In addition to the technologies mentioned above, the Hellenistic period also saw major developments in military technologies – perhaps those in which the rulers were most interested, which explains why they pursued a cultural policy so open to scientists and technologists in general – and those related to architecture, where the emphasis turned to the colossal. A similar tendency to create works of very large dimensions, which required the development of new technologies due to their size, took place in the artistic field and in particular in sculpture. These works of gigantic dimensions were then included in the list of the 'wonders of the world', compiled in the same period (the lighthouse of Alexandria, the colossus of Rhodes).

3.6 THE END OF HELLENISTIC SCIENCE

When the Greeks were expelled from Alexandria, the scientific flowering of the Hellenistic period came to an end. Apart from Posidonius, who worked in Rhodes in the first century BC and who gave a systematic form to the theory of tides, there is no trace of scientists making original contributions in the subsequent period.

An approximate chronology of scientists and authors, who dedicated part of their literary production to science or technology and lived after the expulsion of the Greeks from Alexandria, can be found in the Appendix in Figure A.2. Again, the exact date of their birth or death is not known for some of them, while for others, such as Heron of Alessandria, we are not even sure in which century they lived. Some, like Heron or Claudius Ptolemy, worked in Alexandria, while others, like Pliny the Elder or Vitruvius, were Romans and worked in Rome. Cicero and many others dedicated only a small part of their work to science or technology, and many simply wrote about science without being scientists in the true sense of the term.

As already stated, in this period there was a general neglect of scientific method and a return to natural philosophy, with a strong re-evaluation of the philosophers of the classical period, particularly Plato and Aristotle.

It is very difficult to understand the reasons for the decrease in interest in the quantitative study of transformations and of motion, and the revival of the classical, and more general, themes of natural philosophy. How much this is due to a specific event, such as the expulsion of the Greeks from Alexandria, is also unclear. The changes did not lead to the destruction of the library or of any other scientific or cultural institution, nor even to a specific shift in the cultural policy of the Ptolemies, but only to a change in the managerial frameworks of these institutions. Of course, the original Greek scientists were forced to resettle in other places, such as Pergamon or Rhodes, or in the East, in particular Seleucia on the Tigris. These were all centers ruled by kings who were well disposed towards culture and science and in which there were renowned cultural institutions. However, some hypotheses can be made and possible causes identified:

- The spread of religious cults from the East or from Egypt, with their irrationalistic background. An example is the cult of Isis, which became more and more widespread in Rome. At the same time, there was a rise in pseudo-sciences such as astrology, which was grafted onto the work of Hellenistic astronomy. In this sense it seems that the usual view, which stated that astronomy was a scientific development of astrology, gave way to the observation that astrology was a regression from astronomy towards pseudo-science. Another pseudo-science which gained a following in this period was alchemy, grafted onto the solid foundations of metallurgy and the science of materials, which had allowed progress in metalworking with the casting of large bronze statues and the development of hydraulic machines (for the esoteric aspects of alchemy see Section 11.1).
- The very small number of scientists, even in the period of maximum development of Hellenistic science. It is likely that the works of Euclid, Archimedes, Ctesibius and the doctors, who had positioned medicine on a scientific basis, did not become widespread, as demonstrated by the fact that almost all of them were lost in the subsequent period.
- The lack of dissemination of the scientific culture, even among the few who had access to libraries and were in possession of some culture in a general sense. The scientific method was not understood, and a holistic approach was preferred that gave general explanations of reality, as was typical of natural philosophy.
- The ever-increasing power of Rome, which gradually conquered most of the states in which the Hellenistic culture had spread, apart from the easternmost ones. It was in these eastern states that the works of Hellenistic science survived, where they were first translated into local languages and later into Arabic, after the Arab conquest.

As far as technology is concerned, it was during this period that the attitude of contempt towards those who dealt with these 'servile' things was consolidated, an attitude clearly expressed by Seneca in his *Epistulae morales ad Lucilium* (XC, 13):

"It is well known that certain things date back to our times, such as the use of glass for windows that allow light to pass through the transparent material, or

3.6 The end of Hellenistic science

raised systems for bathrooms and pipes hidden in walls that distribute heat evenly ... these are all inventions of vulgar slaves. Wisdom is located higher up and teaches to the mind and not to the hands".

Where there was no contempt, there was misunderstanding. Vitruvius, who left us a documented treatise on architecture and was certainly the one who had the best understanding of technical issues among the Latin authors, showed that he was unable to understand the scientific texts of Archimedes when the latter dealt with theoretical subjects and particularly when models based on simplified hypotheses were used.

The texts by the scientists, who had lived no more than two centuries earlier, were reported with errors and misunderstandings and were sometimes modified, as was even the case with Euclid's *Elements*.

The only important astronomer of this period, Claudius Ptolemy (about 100 – 175 AD), was no longer able to understand the heliocentric system and returned to geocentrism, which he systemized in a very accurate way. However, even within the geocentric system, he was no longer able to understand that the epicycles were conceived as a mathematical model to describe the motion of planets (a kind of '*ante litteram*' Fourier series expansion) and interpreted them as the motion of the 'crystalline' spheres that carried the planets. The celestial spheres were once again perceived as material objects, even if invisible, and the common nature of matter between the celestial and the sublunary bodies, and above all of the mutual gravitational attraction between all the bodies, was forgotten. The Earth was once again considered to be motionless, not only with respect to its path around the Sun and, even more difficult to understand, around the center of mass of the Earth-Moon system, but also for the daily rotation about its axis. The Earth (still spherical) again became the fixed center of the Universe, with the sphere of the fixed stars (a 'material' object, even if it was made from different matter to that of earthly things) revolving around it once every day.

This brought things back to natural philosophy, in its Aristotelian version, and what once again highlights the neglect of scientific method is that Ptolemy was the first astronomer who also dealt with astrology, a mixture of science and pseudo-science that would last for almost two millennia.

The same thing happened with all the other sciences. Apart from the heliocentric system, which was perhaps too counter-intuitive and difficult to understand at this time, the results of Hellenistic science were not lost, but, as described by Lucio Russo, remained as a "*fossil knowledge*"; that is, knowledge that was repeated in books by the commentators of classical texts, but which were not actually used and, in a sense, remained at the theoretical level [26].

A classic example is the knowledge that the Earth is a sphere, which was reiterated by almost all medieval authors (and was also the basis of the *Divine Comedy* by Dante Alighieri), and yet did not influence the medieval cartographers, who

54 Greek rationality

represented it as flat in their maps – perhaps without even realizing the problem. It is possible they could not do otherwise, since the mathematical techniques introduced by Eratosthenes and other Alexandrian scientists, who had dealt with geography to project the spherical surface of the Earth on a plane, had been lost.

3.7 A HELLENISTIC INDUSTRIAL REVOLUTION?

While Hellenistic science was short-lived and, in the end, left few traces, to such an extent that its existence was often deliberately erased from the history of science, the technology this science produced remained very much alive in the subsequent period, becoming a factor in the increasing economic growth throughout the Roman Empire. Several of the devices described previously (the water pump and the hydraulic organ, the cochlea and the screw, the water wheel, etc.) continued to be produced and used for centuries. An example of the large-scale use of this technology can be seen in the water pumping system of the Rio Tinto copper mines in Andalusia.

In this case, the water was raised by means of a number of hydraulic wheels, probably driven by treadmills and then directly by slaves or animals. The eight pairs of wheels, located in underground tunnels (Fig. 3.2), raised the water by almost 30 meters.

In general, the technologies developed during the Hellenistic period continued to spread, although there seems to have been no further progress. Given their enormous economic importance, and the birth of actual industries based on them, can we speak of a Hellenistic industrial revolution, as we speak of the Industrial Revolution that developed a millennium and a half later in Britain?

There is no doubt that there are analogies. Even in Roman times there was technological development that helped to establish a large number of industries, from the mining industry to the food industry (in particular for the production of oil), together with significant developments in agriculture achieved by applying the results of Alexandrian science to the cultivation of the fields and the breeding of livestock.

However, the market and even an embryonic form of the "consumer society" were almost entirely absent. According to Thomas Pekáry, the purchasing power of the city population of Ptolemaic Egypt was minimal, while the small size of the homes of common people also physically prevented them purchasing goods beyond the needs of subsistence [27]. After all, free workers were almost always paid day by day, without a long-term contract, confirming the fact that job demand always exceeded supply.

In the absence of any form of large-scale production to satisfy private consumers, the operation of the industrial sector mostly remained limited to military, or at

3.7 A Hellenistic industrial revolution? 55

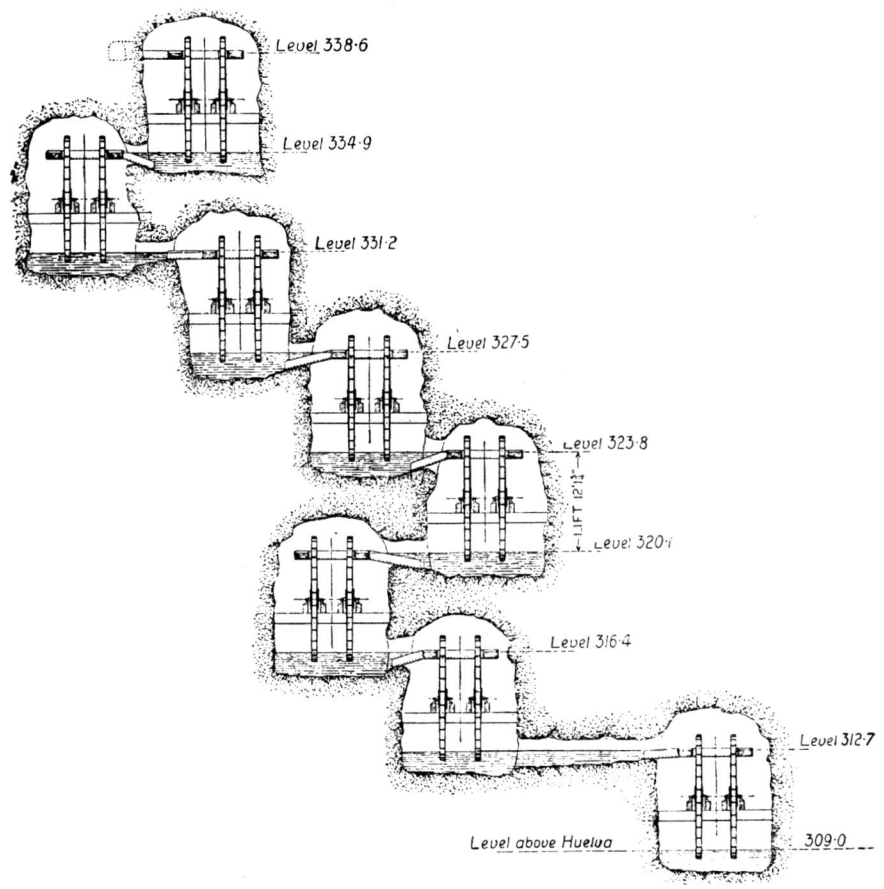

Figure 3.2. Water-lifting system used at the copper mines of Rio Tinto, Spain, in Roman times. [Image from https://upload.wikimedia.org/wikipedia/commons/b/b0/WaterwheelsSp.jpg.]

least governmental, supplies. Many production sectors, such as papyrus processing, were subject to a monopoly.

This economic system meant that even in the period of its maximum expansion, the Hellenistic productive system never reached industrial dimensions. Even the most important orders, such as those concerning the production of weapons and armor for the approximately 300,000 soldiers of the Ptolemaic army, were split between a large number of small shops. This has been confirmed by the archaeological evidence, which shows the total absence of factories.

To prevent the possibility of industrial growth of the shops, there was also the choice by the banks not to grant credit to this type of activity, although it was potentially very profitable. That choice was motivated almost exclusively by prejudice of a cultural nature. Ultimately, given the absence of an adequate network of financial services and of a capitalist production approach, specialists are reluctant to define the Hellenistic technological growth as a true industrial revolution.

If anything, the period can be defined as a two-speed development. On the one hand, there was a series of disruptive technological innovations, the result of a unique rational approach to nature. On the other, there was the archaic economic model, not yet able to develop the potential of these conquests fully.

References

1. Stark R., *How the West Won. The Neglected Story of the Triumph of Modernity*, Intercollegiate Studies Institute, Wilmington, Delaware 2014.
2. *Ibid*, quoting Martin L. West.
3. Jaynes J., *The Origin of Consciousness in the Breakdown of the Bicameral Mind*, Houghton Mifflin, Boston and New York, 1976.
4. Anaxagoras, Diels-Kranz fragment 49 B 12
5. *Ibid*, nt 23 B 57.
6. Plato, *Protagoras*, 320c–322d
7. Xenophanes, Diels-Kranz fragment 21 B 18.
8. Sophocles, *Antigones,* v. 332 and following.
9. Cambiano G., *Platone e le tecniche*, Laterza, Rome-Bari, Italy 1991, p. 55
10. Democritus, Diels-Kranz fragment B 154.
11. Herodotus, *Historiae*, I, 197
12. *Ibid*, II, 84.
13. *Ibid*, III, 129
14. *Ibid*, II, 77
15. Hippocratic Corpus, *On the sacred disease*, chapters 1-2.
16. *Ibid*, chapter 14.
17. Hippocratic Corpus, *The Arts*, chapter 3.
18. Reference 9, p. 37.
19. Reference 17, chapter 12.
20. Hippocratic Corpus, *On Ancient Medicine*, I, chapter 1
21. *Ibid*, chapter 2.
22. Thucydides, *The Peloponnesian War*, II, 47.
23. *Ibid*, II, 51.
24. Pliny the Elder, *Natural history*, VII, 37.
25. Russo L., *The Forgotten Revolution*, Springer, Berlin 2004.
26. Russo L., *Flussi e riflussi*, Feltrinelli, Milan, Italy, 2003
27. Pekàry T., *Die Wirtschaft der griechisch-römischen Antike*, Steiner, Wiesbaden 1979.

4

From Abraham to Jesus: The Judeo-Christian rational horizon

4.1 IN THE BEGINNING WAS THE *LOGOS*...

The encounter between the rationality of the Greek philosophers and the Olympic religion immediately gave rise to a severe conflict. The attempt to claim for humankind the ability to know and describe reality using only the human intellect (*nous*) aroused scandal and accusations of impiety, as evidenced by the trials that led to the exile of Anaxagoras and the death of Socrates.

This was somewhat inevitable. The ancient Greek religion was based on a polytheistic and deeply anthropomorphic conception of the gods, who frequently intervened in everyday life involving humans in their continuous controversies. The line between nature and the supernatural was almost non-existent and the gods used to come down from Olympus to act on Earth, even in disguise. Humans had to know their place and behave in an ethically correct manner, and even then, the reward was far from being assured. The gods were very volatile and only concerned themselves with human well-being if doing so did not conflict with their own interests.

Faced with this, attempting to describe reality or to understand its workings with only the help of the human mind was therefore as unfeasible as it was useless. Consequently, in order to establish scientific thought, it was necessary to deny the existence of the gods of Olympus and replace them with a single divine ordering principle of the universe, an omniscient and perfect cosmic Mind. Only in this way could a homogenous relationship be established between the human mind, the divine mind and its creation, that is, the universe itself.

The first philosopher to envisage such a concept, between the sixth and fifth centuries, was Heraclitus of Ephesus (535 BC – 475 BC). Unfortunately, however, the works of a philosopher who in all probability was one of the greatest intellectuals of the ancient world were completely lost save for a few fragments. In addition to his famous observations on the continuous change of reality (*"one cannot bathe twice in the same river, and one cannot touch twice a mortal substance in the same state"* [1]), there are also a few short sentences in which the '*arché*', or the constitutive principle that underlies the whole universe, is mentioned.

For his predecessors Thales and Anaximenes, the '*arché*' was simply a physical element, like water or air. Heraclitus, on the other hand, thought of an abstract entity, the *Logos*. What does this mean? In ancient Greek, this term could be taken to mean, "word", "discourse" and "rationality". For Heraclitus, the *Logos* was all three things simultaneously: It was the divine Word, the rational Word that created and ordered the cosmos; it was the discourse that man used to describe reality, his path of scientific exploration; and it was also the rational order inherently present within nature. According to Heraclitus, with all due respect to the irrational gods of Homer, "*nothing happens by chance, but everything according to* Logos *and necessity*" [2].

As a consequence, humans, also endowed with *Logos*, could investigate the causes of everything happening around them. Over the following centuries, Anaxagoras was the first to enhance this concept with his universal intellect (*nous*), followed by Aristotle, the undisputed reference point of all classical culture. Beginning with a series of logical reasoning, in his *Metaphysics*, Aristotle outlined the characteristics that the divinity must possess to be the divinity.

First of all, the divinity had to be unique. It had to be "*thinking thought*", a perfect and motionless mind that reflected on itself [3]. Moreover, since the universe was governed by the law of movement, the divinity necessarily had to be a "*motionless motor*", an entity that governed and made physical laws work while at the same time not being subjected to them [4].

That Aristotle was considered credible marked a turning point in the culture of the time and in light of his thinking in the Hellenistic age, the ancient thinking of Heraclitus was rediscovered. The leading philosophical trend of this time was Stoic philosophy, one of the major currents of thought of antiquity which was also destined to spread across the Roman world. The pillar of Stoicism was the idea of a universal *Logos*. As observed by Antony Long, according to this school of thought "*Nature is not merely a physical principle (...), but it is also something that is endowed with rationality par excellence. What holds the world together is a rational supreme being, God, who governs all events for purposes that are necessarily good*" [5].

Despite physiological attempts at identification, the gap between this God and the Zeus of traditional Greek religion was increasingly marked and difficult to fill. Consider also that, thanks to the conquests of Alexander the Great, Greek philosophy had come into contact with other civilizations and religions of the Near East which, to a higher degree, fulfilled the growing need of a deity – *Logos*.

The most fortunate and prolific synergy above all was with Judaism, because of the natural convergence between Greek philosophical concepts and what Rodney Stark calls "*Jerusalem's rational God*" [6]. There were many similarities between the monotheistic deity of the Hebrew Bible and the *Logos*, to such an extent that the two entities would soon inevitably coincide.

The best proof of this convergence is undoubtedly the beginning of the Gospel of John, in which the myth of creation is rewritten in philosophical terms: "*In the beginning was the* Logos*, and the* Logos *was with God, and the* Logos *was God. He was in the beginning with God. All things were made through Him, and without Him nothing was made that was made*" [7]. The Latin '*Vulgata*', the most widespread translation, uses the term "*Verbum*". However, we should keep in mind that *Logos* is also rationality, and that "*all things were made through Him*". For the first time, philosophy and religion seemed to speak the same language.

4.2 THE RATIONAL GOD OF THE OLD TESTAMENT

At first glance, defining the Yahweh of the Hebrew Bible as a "rational God" may seem like a provocation. How does this idea reconcile with the anthropomorphic and non-omniscient God, who "walks in the garden" of Eden? [8] Or with the distant and unknowable divinity of the books of Job and Ecclesiastes?

Any attempt to reduce the Bible – or Tanakh – to an organic and consistent religious system is an operation that easily lends itself to many objections. Whatever the point of view adopted, there are many passages that differ, diverge from or simply oppose that view.

As the Hebrew name Tanakh[1] suggests, from a historical point of view the Old Testament is a heterogeneous collection of works written in different historical periods, reflecting different concepts of the universe. It is often possible to identify different authors and philosophies even within the same book. This is the case, for example, with Genesis and Isaiah, whose content has been attributed to four and three different authors respectively[2].

---

[1] The Hebrew word *TNK*, spoken as Tanakh, is actually an acronym for *Torah* (Instruction), *Neviim* (Prophets) and *Ketuvim* (Writings) which synthesizes in a single term the three basic parts of the Old Testament.

[2] According to the thesis of Julius Wellhausen, Genesis was composed of two independent sources, called "J" and "H". One indication of this dualism, by way of example, are the two distinct and autonomous stories of the creation of man and of the animal world in Genesis 1 and 2, and the presence of sections of the book in which God is referred to only as Yahweh, or only

Beyond the historical debate, we must disregard these differences and focus on the dominant Jewish concept, with which Greek philosophers came into contact in the third century.

In this perspective, there are essentially three fundaments of the biblical universe: the uniqueness of God; the perfect rationality of his creation; and the possibility for humans to understand the working of the earthly world and to govern it with their own instruments. From the beginning, humans are given the opportunity to use nature to their advantage. In Genesis, God orders the man and the woman: *"Be fruitful and increase in number; fill the earth and subdue it. Rule over the fish in the sea and the birds in the sky and over every living creature that moves on the ground* [9]. The purpose of this primacy over other creatures is to allow humans to manipulate nature with the use of agriculture and technology: *"the Lord God took the man and put him in the garden of Eden to tend and keep it"* [10].

Humans are left free to investigate the universe. Having a limited mind, they cannot fully understand the divine plan on a cosmic scale, but are nevertheless able to exercise their intellect to describe the workings of the reality around them, initially by giving all things a name: *"Out of the ground the Lord God formed every beast of the field and every bird of the air, and brought them to Adam to see what he would call them. And whatever Adam called each living creature, that was its name"* [11].

Unlike the ancient Greek tradition, which considered the names of the inventors unworthy of being remembered, Genesis notes in detail the names and genealogies of the ancient patriarchs, to which the first technological innovations were attributed:

"Adah bore Jabal: he was the father of those who dwell in tents and have livestock. His brother's name was Jubal: he was the father of all those who play the harp and flute. And as for Zilla, she also bore Tubal-Cain, an instructor of every craftsman in bronze and iron" [12].

The superiority of humans over the other creatures is sanctioned by the possession of a rational mind, represented by the divine Spirit (*ruah*) blown into their body at the moment when man was created from clay. Unlike Prometheus, the Jewish deity does not give technology to man, but simply intellectual ability.

Humans could then make use of rationality to develop new techniques, such as jewelry and precision craftsmanship. As God says to Moses in the Book of Exodus, *"I called by name Bezaleel, the son of Uri, the son of Hur, of the tribe of Judah and*

---

with the name Elohim. The two sources could be ascribed respectively to the Southern and Northern Kingdoms during the period of division of the Jewish monarchy (930–721 BC). To these would be added the third and fourth hand of a later editor, who put the story together or rationalized it. The book of Isaiah, in contrast, is conventionally distinguished between the first Isaiah and the subsequent "deutero-Isaia" and "trito-Isaia".

4.2 The rational God of the Old Testament

I have filled him with the Spirit of God, in wisdom, in understanding, in knowledge, and in all manner of workmanship, to design artistic works, to work in gold, in silver, in bronze, in cutting jewels for setting, in carving wood, and to work in all manner of workmanship" **[13]**. Here, the Spirit essentially performs a function similar to that of the *Logos* of the Stoics.

Gradually, the lexicon of the Bible identifies this rationality which acts as a bridge between God and humans in a more precise way. Initially, in the Hebrew version, the intellect was simply identified with the heart (*lev*), the organ in which individual conscience was thought to reside[3]. However, when it was necessary to distinguish the purely rational sphere from that of the affections, since the heart was also the seat of feelings, the soul (*nefesh*) was used. Curiously, this was thought to reside in the kidneys[4]. The link between the intellect and the soul gradually strengthened, to the point that, in modern Hebrew, *nefesh* means "mind".

The major change, however, came from the encounter between Judaism and Greek philosophy, which occurred following the conquest by Alexander the Great. Culturally, the most fertile terrain was that of Ptolemaic Egypt. The Jews of Alexandria accepted the Hellenistic culture much more quickly than their Palestinian compatriots, to the point that it soon became necessary to translate the Bible into the Greek language. The integration had been so deep that they abandoned their linguistic identity and kept only their religious identity.

In the most widespread and authoritative Greek translation to describe the rational sphere, the so-called version of the Seventy, the simplistic association with the body's organs was abandoned in favor of the far more specialized language of philosophy. This is particularly evident in the biblical books written more recently, in which we can distinguish between rational soul (*psyche*[5]) intellect (*nous*) which is understood as a specific function, and thought (*dianoia*) which is the idea produced by reasoning[6].

---

[3] In Proverbs 15, 14 for instance, it is said that "*The heart of him who has understanding seeks knowledge*", while in Deuteronomy 29, 4, Moses explains to the Jews that "*yet the Lord has not given you a heart to perceive*".

[4] The distinction can be found in Jeremiah 11, 20: "*But, o Lord of hosts, you who judge righteously, testing the kidneys (mind) and the heart*"; in Psalms 7, 9: "*Oh, let the wickedness of the wicked come to an end, but establish the just; For the righteous God tests the hearts and the kidneys (mind)*"; and in Psalms 25, 2: "*Examine me, O Lord, and prove me; Try my kidneys (mind) and my heart*".

[5] From this term comes the modern idea of "psyche".

[6] For instance, the word *psyche* occurs in Chronicles 9, 1, while the Book of Wisdom make reference to the ideas of *nous* (4, 12) and *dianoia* (4, 14). The term *Logos* instead retains the basic meaning of "word", without any association to the concept of Universal Reason. This new meaning will be introduced in the Gospels.

These traits, described in an ever more accurate way, are attributed to both mortals and Yahweh. It is the possession of a rational mind that establishes a link between humans and the divinity who created them in his own image and likeness.

4.3 JUDAISM AND CENTRIFUGAL THRUSTS

The historic phase, which began in the third century BC and continued for several centuries, is called Judaism[7]. In this era, Jewish rationality also extended to ethics, assuming a particularly radical form. Under the impulse of the so-called sadocites, the priestly class of Jerusalem that claimed to be descended from Sadoq, the idea of a universal retributive justice was affirmed. According to this concept, every event – favorable or unfavorable to humans – could be rationally explained by ascribing it to two causes. The first was the material one, connected to the world of nature, while the second, of divine origin, was connected to individual behavior. The underlying assumption was that if the universe was rationally ordered, it would also necessarily have to follow the principle of cosmic justice. Therefore, according to Jewish thinking, negative events would always affect people who didn't respect the moral laws of the universe, leaving no room for chance. If this view was not sufficient to describe reality, the hypothesis of atonement for the sins committed by family members and previous generations was used.

Such a strongly dogmatic concept of the rational ordering of the universe, however, was clearly problematic in its application, and immediately generated lively debates. In a well-known passage from the Gospels, commenting on the tragic death of eighteen people caused by the fall of the tower of Siloam, Jesus distances himself from those who claim that the victims died because of their sins [14]. The concept is also repeated with the healing of a blind man: that the view of the sadocites is an excess of rationalism that clashes with reality [15].

Three centuries before Jesus, many voices had already spoken against this dominant concept. Some of the more authoritative among those voices also became part of the biblical canon, such as the books of Job and Ecclesiastes.

In these two sapiential books, an apparent return to irrationalism can be found, but only as a polemical and provocative reaction to the doctrines of the sadocites and to the optimism of the book of *Proverbs*, which reflected this automatic association between individual behavior and destiny on several occasions, ignoring the problem of evil and injustice.

---

[7] For further information about the Judaism of the Second Temple, the Zadokides and the widespread Jewish centrifugal thrusts, see Boccaccini G., *Oltre l'ipotesi essenica: lo scisma tra Qumran e il giudaismo enochico*, Morcelliana, Brescia 2003; and Sacchi P., *The History of the Second Temple Period*, T. & T. Clark, Edinburgh 2004.

4.3 Judaism and centrifugal thrusts

The episode in which Job addresses God, unjustly plagued by a number of misfortunes of supernatural origin and lamenting the lack of respect for cosmic justice toward him, is emblematic. The divine response is very harsh, and denies humankind any ability to understand the workings of the reality surrounding them: "*Where were you when I laid the foundations of the earth? Tell me, if you have understanding. Who determined its measurements? Surely you know! Or who stretched the line upon it? (...) Who can number the clouds by wisdom? Or who can pour out the bottles of heaven, when the dust hardens in clumps, and the clods cling together?*" [16].

God does not limit himself to stating that it is impossible for humankind to know the ethical motivations of what happens, but he also denies all their scientific ambitions. In fact, Yahweh sketches a long description of the animal and vegetable kingdom, of celestial and space physics, and reveals to Job that even these realities cannot be fully understood by the human mind. A rational order of the universe exists, but the divine providence which rules it is governed by laws which do not coincide with human laws. The irrationalists who deny the existence of this equilibrium are wrong and should limit themselves to a respectful silence. Job replies: "*Behold, I am vile. What shall I answer You? I lay my hand over my mouth. Once I have spoken, but I will not answer; yes, twice, but I will proceed no further*" [17].

The same limitation is expressed in the book of Ecclesiastes, in a particularly pessimistic form: "*For in much wisdom is much grief, and he who increases knowledge increases sorrow*" [18]. The implication for scientific research is particularly heavy: "*I communed with my heart, saying, 'Look, I have attained greatness, and have gained more wisdom than all who were before me in Jerusalem. My heart has understood great wisdom and knowledge'. And I set my heart to know wisdom and to know madness and folly. I perceived that this also is grasping for the wind*" [19].

The human ambition to know the world is bound to be frustrated because humans are too limited to be able to understand what surrounds them, and must therefore limit themselves to the fear of God [20].

Once again, however, the existence of a cosmic order is not in doubt, a concept that actually underlies the first chapter of Qoeleth: "*One generation passes away, and another generation comes; but the earth abides forever. The sun also rises, and the sun goes down, and hastens to the place where it arose. The wind goes toward the south, and turns around to the north; the wind whirls about continually, and comes again on its circuit. All the rivers run into the sea, yet the sea is not full; to the place from which the rivers come, there they return again. All things are full of labor; man cannot express it. The eye is not satisfied with seeing, nor the ear filled with hearing. That which has been is what will be, that which is done is what will be done, and there is nothing new under the sun. Is there anything of which it may be said, 'See, this is new'? It has already been in ancient times before us"* **[21]**. Once again, however, man cannot understand the functioning of this rational universe. In truth, however, the anti-Sadocite controversy mainly concerned the field of ethics and the impossibility of knowing the destiny of man, dealing only marginally with the field of sciences.

Moreover, these critical positions did not belong to the dominant Jewish thinking and must be traced back to their specific historical dimension, i.e., to the centrifugal thrusts that, between the third century BC and the first century AD, induced many peripheral realities. This even included refuting the central authority of the sadocite priests of the Temple of Jerusalem because of their excessive proximity to the Hellenistic world. Another such group was the apocalyptic community of Qumran, who wrote the famous Dead Sea Scrolls and claimed to be descended from a charismatic "Master of Justice", in open conflict with the spiritual leaders of Jerusalem.

Regardless, these are essentially minority realities. The dominant Judaic culture welcomed the influence of the Greek world, particularly in terms of universal rationalism. According to historian Morton Smith, *"Hellenization even reached the basic structure of rabbinical thought"* **[22]**. In this sense, a particularly significant example comes from the works of Philo of Alexandria, a Greek-speaking Jewish philosopher who proposed a perfect synthesis of the Old Testament with the thinking of Plato. For the first time, religion and philosophy were joined by rationality. For Philo, God was *"the pure and perfect Mind of the universe"* **[23]**.

Even in the context of the literature of wisdom, the conflict between the book of Proverbs and those of Job and Qoeleth is superseded by later books which are more influenced by the Greek-Hellenistic philosophy, such as Sirach and Wisdom.

According to these writings, the true wise man is one who does not claim to bring the balance of nature forcibly back into the realm of human law, and recognizes that its functioning comes exclusively from God. At the same time, however, he can master all the physical sciences.

Sirach, referring to Job, affirms that humans do not have to worry about the cosmic order or trying to replace the divine providence, which has already ordered the world in the best possible way at the moment of creation. At the same time, however, they do have the moral duty to study its working, in order to get closer to God: *"Listen to me, my son, and acquire knowledge, and pay close attention to my words. I will impart instruction by weight, and declare knowledge accurately. The works of the Lord have existed from the beginning by his creation, and when he made them, he determined their divisions. He arranged his works in an eternal order, and their dominion for all generations; they neither hunger nor grow weary, and they do not cease from their labors"* **[24]**.

The cognitive and exploratory limits of Job and Qoeleth are now outdated, and this type of investigation of reality establishes a strong bond between man and God, since both share a rational intellect. According to the Book of Wisdom, it is a gift that Yahweh bestows upon the scholar to investigate the secrets of nature: "*May God grant that I speak with judgment and have thoughts worthy of what I have received, for He is the guide even of wisdom and the corrector of the wise. For both we and our words are in His hand, as are all understanding and skill in crafts. For it is He who gave me unerring knowledge of what exists, to know the structure of the world and the activity of the elements; the beginning and end and middle of times, the alternations of the solstices and the changes of the seasons, the cycles of the year and the constellations of the stars, the natures of animals and the tempers of wild beasts, the powers of spirits and the reasonings of men, the varieties of plants and the virtues of roots; I learned both what is secret and what is manifest, for wisdom, the fashioner of all things, taught me*" [25].

All the sciences of the Greek world – cosmology, astronomy, zoology, botany and meteorology – were a stage of the journey towards God. Moreover, as Mazzinghi observed, in this passage humankind is not "*denied the procedure typical of sciences*" by virtue of a supposed difference between the mind of man and that of the Creator, but that scientific method is only "*re-evaluated in its origin and its purpose*" [26]. Obviously, from an ethical point of view, the understanding of the learned "*does not contradict the sciences, but overtakes them in penetration*" [27].

Reasoning in these terms, however, one might think that the contribution provided by Judaism to the birth of the Western scientific world is limited to having accepted Greek natural philosophy in the sphere of the sacred, without having provided any original contribution. Nothing could be further from the truth. Giving shape to the world we know above all, and marking a turning point with respect to the classical Greek world, were the aspects of Judaism which are most peculiar and least reconcilable with Greek philosophy.

The rational god of Aristotle and of Hellenistic philosophy, for example, was a distant and impersonal principle who guaranteed cosmic equilibrium. Love, perceived as a necessity that sprang from the lack of something, was completely foreign to him. The Greek ideal of contemplative perfection was to emerge from this particularly cold theology. According to the philosopher Plotinus (204–270 AD), creation was produced by divinity without any intention by passive emanation, just as light is emitted by the Sun. No action, movement or concrete thinking could be attributed to the God-*Logos* because even one such voluntary intervention would have compromised its perfection.

The demiurge, the cosmic craftsman of whom Plato spoke in his *Timaeus*, was to shape and actively construct the earthly world. However, this was not at all a divine being; indeed this otherworldly architect was directly responsible for the imperfection of the reality in which we live. The fervent Greek cultural bias against work and applied science derived from this, as has been previously observed. If divine perfection existed in abstract contemplation, then humans must also strive in this direction, abstaining as much as possible from any manual or practical activity and, above all, not turning their noble intellect to these unworthy activities.

In contrast, the biblical God intentionally designed and constructed every component of the universe, and is not bound by a state of contemplative isolation. Animated by love for humankind, he acts and intervenes constantly in history so that his rational project finds its fulfilment. In the first chapter of Genesis, in particular, God works actively on His creation for six days. It is a productive process in all respects, followed by a dissipation of energy and the institution of a day of rest. In this way, from the first chapter, the Hebrew Bible draws a clear parallel between the work of humans and that of God, strongly re-evaluating the role of those who perform a manual activity. In short, there is nothing wrong or inferior in the work of the craftsman, since Yahweh himself *"has his hands dirty from work"* using the material with which the cosmos was constructed. If these occupations were typically servile and unworthy of free men to the Greeks, in the Bible they represent an imitation of the divine work, paving the way for the idea of an applied science.

4.4 CIRCULAR TIME, THE MYTH OF THE AGES AND PROGRESS

The revolutionary innovations introduced by Judaism are not limited to the concept of divinity and the role of manual labor. A particularly significant aspect, which underlies the notion of technological progress, is represented by the perception of the flow of time.

From a geometric point of view, the circle is the perfect figure, without a well-defined beginning or arrival point. In the same way, in line with the thought of Eastern civilizations, many Greek philosophers maintained that time was a cycle bound to repeat itself forever, starting from the abstract symbol of the circle as a perfect geometric figure. For example, in *Timaeus*, Plato states that the demiurge, realizing the universe, *"assigned it a movement that suits its body, and that is the one that mostly concerns intelligence and thought. Therefore, making it turn around in the same way, at the same point and in itself, he made it to move with a circular motion"* **[28]**. This movement characterizes both the orbit of our planet from the spatial point of view as well as the course of history on the temporal level. Everything is a perfect circle. Moreover, human lives would also have been perceived as eternally circular, as we learn at the end of *The Republic* from the

4.4 Circular time, the myth of the ages and progress

myth of Er [29]. In fact, Plato took from Pythagoras the theory of metempsychosis, or the reincarnation of the soul into a new individual.

Human life, in this way, was seen as an eternal repetition of previous lives, in the same way as the day-night cycle or the cycle of the seasons. In short, cyclic repetition would represent the immutable space-time order of the universe.

This concept was further developed by Stoic philosophers, who, as mentioned, introduced a concept based on the rational order of the universe governed by the God-*Logos*. According to them, this rational order was based on the eternal return; that is, on the cyclic succession of the same events. Each time, the end of the cycle of time was marked by the *apokatástasis*, a universal conflagration of purifying fire that brought everything back to its primordial condition.

The esoteric symbol of the *Ouroboros*, the cosmic serpent that eats its tail in an eternal repetition of itself, was derived from these doctrines. This representation, which would also disseminate among the Gnostic sects and in medieval esotericism, condenses the ideal of cyclic perfection.

Figure 4.1. Late-medieval representation of the Ouroboros, drawn by Theodoros Pelecanos in the alchemical treatise Synosius. [Image from https://upload.wikimedia.org/wikipedia/commons/7/71/Serpiente_alquimica.jpg.]

For those who accepted this view of the world, the cycles which follow each other did not mean just a simple analogy between present and past, from which one could at least derive the need to study history to understand past errors in order to avoid repeating them. The cycles were in fact perfectly isolated from each other by the cosmic combustion, a cataclysmic event that destroyed every form of memory and made any progress or learning impossible with respect to the previous time. Above all, it was not a simple analogy between present and past, but a complete and total repetition of the same events, an aspect that prevented any form of innovation.

The phrase *"nothing new under the sun"* effectively summarizes the biblical book of Ecclesiastes which, also with regard to time, marks a clear discontinuity with respect to official Judaism and stresses the deep pessimism inherent to this concept: *"Is there anything of which it may be said, 'See, this is new'? It has already been in ancient times before us. There is no remembrance of former things, nor will there be any remembrance of things that are to come by those who will come after"* [30].

The concept of cyclical time is antithetical to modern science, unless we hypothesize a cycle of a duration (tens or hundreds of billions years) sufficient to coincide with the fluctuations of a possible, and hypothetical, oscillating universe, which expands after the Big Bang and then, after reaching its maximum expansion, contracts into what is sometimes called a Big Crunch, before starting over again. Between one cycle and the next there would be a singularity that erases all traces of the past, the modern equivalent to the *apokatástasis*.

In contrast, Judaism is based on a linear concept of time. According to the Bible, the world was created at a precise moment, beginning a historical process which is advancing in a straight line and involves both the world and humankind. This path will end in another well-defined moment in which the world will cease its existence, although there are different interpretations about how this will happen and the consequences of the end of the world. Regardless, time is perceived as a rectilinear segment that coincides with the limited existence of the earthly world.

The concept of linear time is a Judeo-Christian creation in all respects, which led to a fundamental change in understanding the reality that surrounds us. In fact, this vision of the world is open to innovation and scientific discovery, since it excludes any form of repetition. However, the notion of linear time is not in itself sufficient to imply the idea of scientific progress automatically, even though it constitutes an essential prerequisite for it.

In the biblical account, the creation of humankind is followed by its expulsion from Eden and the loss of immortality. Later on, as a result of the incest among women and the *"Sons of God"* [31], i.e., the fallen angels, humans also lose longevity, seeing their own lifespan shrink to 120 years. Initially, humans did not

4.4 Circular time, the myth of the ages and progress

even have to work hard to find their livelihood, being sentenced to *"cultivate the land with the sweat of his forehead"* following the Fall of Adam.

Work and technology had already existed previously, and were distinctive features of the human being in the Bible, but it is with the introduction of pain and suffering that the use of tools actually becomes necessary. Innovations, therefore, do not involve any improvement.

According to the more apocalyptic views, the only positive improvement will be represented, in an indefinite future, by the return of the two prophets Enoch and Elia who had ascended to heaven, or by the coming of the Messiah, a figure who will either restore the power of the sadocite or bring David's family back to the throne. However, in the Jewish context, it will be a political-spiritual redemption that has little to do with the Fall of Adam and with the other 'universal defeats' suffered by humankind, which are and remain irreversible.

In short, the path of history is a degenerative one, a continuous regression from a situation of primordial perfection. In this, albeit to a lesser degree, Judaism is still deeply conditioned by the concept, very widespread in antiquity, which exists in the so-called Myth of the Ages.

This is an idea which mainly characterizes the Greek world. From the Platonic myth of Atlantis[8] to the Hesiod legend, there are many testimonies of a Golden Age[9] in which humans lived in peace with each other and with nature, with no need for technology. For the Romans, it was the heavenly age of Saturn[10]. For those who believed in this interpretation, machines were not a conquest but only a "necessary evil". They did not provide progress for the state of humankind, but rather adaptation required by the involution to which it was subjected with the passing of the ages.

Moreover, as mentioned, Greek thinking was marked by a fundamental dichotomy. On the one hand, the philosopher was the theoretician of scientific progress, and asked questions about the world and nature with the aim of increasing his knowledge of the reality surrounding him. On the other hand, the philosopher himself indulged in fantasies concerning a mythical 'golden' past in which none of this was needed.

---

[8] Plato describes this ancient and very advanced civilization in the dialogues *Timaeus* and *Critias*.

[9] In his *Works and Days*, the poet Hesiod (seventh century BC) described an age of primeval happiness. In this time, Earth was populated by a *"Golden lineage of mortal men"*. Hesiod then described the following four ages in chronological order: The Silver Age, the Bronze Age, the Age of Heroes and the Iron Age. Every new age came with a severe regression.

[10] See, for example, the *Bucolics* or *Eclogues* of Virgil.

The utopian illusion of being able to live in a 'natural' state, without any technological support, was a clear symptom of the refusal to accept the role and importance of applied science. The final turning point would come with the advent of Christianity, which would bring to completion the reappraisal of humankind and of the earthly world in which they lived, while also overturning the anti-technological and pessimistic concept of the myth of the ages.

4.5 THE CHRISTIAN DNA OF THE TECHNOLOGICAL WEST

In the contemporary collective imagination, Christianity and technological innovation are two antithetical phenomena. It is a common opinion that the roots of the material progress of the West stemmed from a secular and illuministic substratum, which emerged in the mid-eighteenth century, and that in previous centuries the Christian religion did nothing but curb and hinder the rise of science and innovation by every possible means. This thesis, never demonstrated scientifically, starts with single episodes, such as Galileo's trial, which are systematically considered out of their context to take on a universal and dogmatic dimension.

To speak of 'dogmas' in the secular field may seem provocative, but this is not the case. At a scholastic level, it has become a (bad) habit to re-read history in the light of some unproven milestones, which cannot be ignored. That of the intrinsic opposition between Christianity and technological innovation is one of the strongest, from which other minor dogmas derive, such as the identification of the Middle Ages as the 'Dark Ages'.

One of the objectives of this interpretation, as shown by the American historian Rodney Stark in his works, is to deny the West its present intellectual and technological primacy. In spite of this being obvious and incontrovertible, this uncomfortable status is in fact increasingly being challenged by the spread of politically correct ideology, which strives to understate its uniqueness as much as its identity.

If, instead, we choose to disregard the dogma of the contrast between Christianity and science, it becomes clear how the Gospels and the writings of the Fathers of the Church represent the decisive element in the development of technology and the construction of Western identity. Therefore, separating scientific technology from Christianity effectively means separating the two halves of an essential pair, thus depriving the European and American rationality of its historical identity.

In summary, the first process in the formation of the rational DNA of the West was the Greek stage with the birth of natural science and philosophy, which for the first time superseded magical thought to begin to question the world around us. The second stage was the Jewish one, characterized by the advent of a rational religion capable of establishing a synergy with the thinking of Greek philosophers

4.5 The Christian DNA of the technological West

and, at the same time, overcoming the cyclic perception of time and the contempt for material work.

The last decisive step, however, is represented by the contribution of Christianity, meaning both the historical teaching of Jesus Christ and the subsequent theological elaboration of his figure by the early Christian authors. These three ingredients would meet and combine fully only in the Middle Ages, a period that, in fact, can be seen as the basis of contemporary science.

"*Only Westerners thought that science was possible, that the universe functioned according to rational rules that could be discovered. We owe this belief partly to the ancient Greeks and partly to the unique Judeo-Christian conception of God as a rational creator*", Stark effectively summarizes [6]. David F. Noble also understands the role played by Christianity in the birth of the technological West, even if he provides a negative reading of both elements [32].

As mentioned in the previous section, both classical Greece and the Jewish Bible gave way to a pessimistic 'Myth of the Ages', which traced a regressive path for humankind devoid of any form of progress and relegating technology to a mere form of adaptation to this regressive reality.

With Jesus Christ, the model is completely subverted. With his earthly mission, the Messiah restores the Golden Age, cancelling the Fall from Eden and giving back to humankind its original perfection. According to Noble, "*Recalling the divine likeness of the first Adam, the advent of Christ promised the same destiny for a redeemed mankind*" [32]. This interpretation dates back to Paul of Tarsus, who repeatedly defined Jesus as "*new Adam*", "*second Adam*" and "*last Adam*" in his letters. According to the Fathers of the Church, a first turning point occurred at the time of the Resurrection and with the descent of the Holy Ghost on the first Christian community, even if the definitive return to the state of perfection will occur only with the *parousia*, the second coming of the Messiah.

The concept of history is reversed optimistically. After Jesus, the journey of humankind on this Earth is a path marked by progress and continuous improvement. The return of the Golden Age, however, does not include the definitive abandonment of technology and manual work and a return to the primordial "state of nature" of which the Greek and Latin poets and philosophers were dreaming. On the contrary, if anything it passes precisely through the revaluation of matter and the most humble and despised works. Jesus himself is a craftsman, and turns his preaching to the fishermen and shepherds of Galilee.

Before beginning his own preaching, the Gospel of Mark[11] tells us that Jesus had in fact worked extensively in Nazareth as a *tekton*, a polyvalent term that embraces various craft activities including woodworking and carpentry. Unlike

---
[11] Gospel of Mark 6, 3. The parallel passage of the Gospel of Matthew 13, 55, instead tells us that Jesus was the "son of the *tekton*", Joseph. Apparently, it was the family business.

other Middle Eastern prophets, therefore, Jesus was not at all of poor social background, because the *tekton* was a relatively affluent social category, even if it belonged to the world of manual workers and, as such, was always relegated to a subordinate role.

This identity accompanied Jesus throughout his teaching. His parables are full of images taken from the world of work, ranging from building to business administration and from agriculture to shepherding. Overcoming the Greek dualism that placed spirit and matter in opposition also coincided with the social redemption of the figure of the craftsman and his work. On a theological level, playing a particularly important role in the re-evaluation of the materiality and the earthly world operated by Christianity, was the concept of divine incarnation, in contrast to the unmoved and contemplative god of Aristotle. Christianity speaks of a God who becomes man, who works and lowers himself to live with the humblest people, thus sanctifying both them and manual activities like the work of craftsmen. If the Jewish God had "dirtied his hands" with the act of creation, the Christian God brings this concept to the most extreme consequences. Think of the key episode of the washing of the feet, illustrating that there is no occupation, however servile, which is unworthy before God. As if this were not enough, Jesus concludes his earthly life by returning, through ascension, to the perfection of the spiritual world with his own material body.

In this way, the Greek and Roman prejudice against matter is definitively laid to rest, a prejudice which, as previously mentioned, had prevented the birth of applied science for centuries, with the sole exception of medicine.

At the same time, Christianity elevates humans and their natural talents, in clear contrast to the strong pessimism of the books of Job and Ecclesiastes. As Noble acutely observes, among all the ancient religions, *"Christianity alone blurred the distinction and bridged the divide between the human and the divine. Only here the salvation came to signify the restoration of humanity to its original God-likeness"* [33].

Going beyond the already optimistic creation "in image and likeness" of the Jewish Genesis, Christianity reworks the messianic prophecies of the Book of Daniel by defining Jesus Christ as the "son of Man" and identifying him with both the divinity and the human being [34]. With the Trinitarian doctrine and the figure of the Son, humans, now free of the burden of the original sin, completely find a place within the theological description of the divinity itself. Ultimately, Christianity ennobles the nature of humans and their abilities, recognizing in them the ability to put into practice the will of God and to perform extraordinary actions.

After overcoming the pessimistic myth of the Eras, ending the prejudices against matter and re-evaluating the human being up to the threshold of divinity, all that remained was to start a path of technological innovation. The transition, however, was not yet as obvious as it might seem...

4.6 THE TWO SOULS OF CHRISTIANITY

Hinting at the discordant viewpoints coming from the biblical books of Job and Ecclesiastes (among others), we have seen how the Judaism of the three centuries before Christ was not at all a unitary phenomenon and was instead marked by strong centrifugal thrusts which bore a radically different world view.

A similar situation also affected early Christianity, which in turn was a minority phenomenon compared to Judaism itself. In particular, we can distinguish between an active Christianity, which is based on the principles outlined in the previous chapter, and a contrasting mystical-ascetic trend which inherited the Greco-Roman disdain for the material. This dichotomy emerged immediately. With regard to the earthly world, the New Testament presents a particularly articulated framework and there are many quotes in support of both positions.

While the more humanistic concept of Christianity, which re-evaluates work and the material world, is based on the elaboration of Paul and the early Fathers of the Church, the ascetic component is actually linked to statements by Jesus himself: *"It is easier for a camel to go through the eye of a needle than for a rich man to enter the kingdom of God"* [35]. *"Man shall not live by bread alone"* [36]. *"If anyone comes to Me and does not hate his father and mother, wife and children, brothers and sisters, yes, and his own life also, he cannot be My disciple"* [37]. The list could go on, but it would require a long study in terms of analysis and contextualization.

In addition, beginning with the forty days spent by Jesus in the desert and his struggle against temptation, part of early Christianity is even oriented towards a hermitic existence, completely isolating itself from the world. In imitation of the Aristotelian God, pure spiritual contemplation is privileged, rejecting as far as possible any contamination of a material nature. The expectation of an imminent second coming of Christ is often added to this particularly radical perspective. There is no time for material progress, as this world is about to end.

On the one hand, this branch of early Christianity certainly claims a greater adherence to the most drastic and provocative points of the teaching of Jesus, but on the other hand, it distances itself from the overall view of the teachings of the Gospel, which provides for Christians an existence deeply immersed in the daily world and in matter. Not for nothing, after the forty days of asceticism, did Jesus preach to shepherds, fishermen and peasants for three years through villages and towns.

The passage in the Gospel of John which identifies Judas Iscariot as the treasurer of the group of apostles shows how the circle of Jesus was not averse to money and materiality [38]. Similarly, even the maxim *"render to Caesar the things that are Caesar's, and to God the things that are God's"* evidently does not address an audience of ascetics [39].

However, as already stated, leaving aside the controversial issue of adherence or not to the original message, it is still possible to observe how, within the origins of early Christianity, two opposing interpretations came to the fore: The ascetic soul, which refuses all contact with the earthly world and lives in anticipation of the imminent return of the Messiah; and the soul, which we will call 'humanist' here, that instead postpones Christ's second coming to a not-so-near future and focuses on the revaluation of humans and their work.

Augustine of Hippo (354–430), who, judging by his writings, was clearly in the second category, wondered: *"In addition to the arts of good living and of reaching eternal happiness (...) from human ingenuity, many important arts have been invented and practiced. (...). Human activity has reached marvelous and wonderful realizations of clothing and buildings, has progressed in agriculture and navigation, has conceived and executed works in the production of various ceramics and also in the variety of statues and paintings, has set up in theatres actions and admirable representations for the spectators, incredible for the listeners; used many means to capture, kill and tame wild animals; invented all kinds of poisons, of weapons, of instruments against men themselves, and many medicines and aids to defend and recuperate health (...) exposed with great precision and intelligence the exact knowledge of geometry and arithmetic and the course of the stars and has reached a deep knowledge of physics"* **[40]**.

Moreover, as Paul (5 AD – 67 AD) pointed out in his Second Letter to the Philippians, the apocalypse is not imminent, and in this interim period humans must put their exceptional qualities to good use to improve the world in which they live: *"Now, brethren, concerning the coming of our Lord Jesus Christ and our gathering together to Him, we ask you, not to be soon shaken in mind or troubled, either by spirit or by word or by letter, as if from us, as though the day of Christ had come. Let no one deceive you by any means"* **[41]**.

The Gnostic sects that wrote the apocryphal gospels of Nag Hammadi belong instead to the ascetic soul of Christianity. In truth, even if referring directly to the teaching of Jesus, these currents are only nominally Christian, as the authors of an independent esoteric thought. Their influence on early Christianity, however, was very strong.

With the Gnostics, the contempt for technology and progress reached its peak. The apocryphal gospels brought together the Persian Light-and-Darkness dualism with the philosophical theories of Plato, identifying matter and the earthly world in which we live with absolute evil.

In the opinion of this movement, Jesus entrusted his secret teaching, according to which the earthly world is nothing but a virtual prison, oppressive as well as illusory and built by a false god, to some particularly chosen disciples. It is thus useless devoting one's intellect and energies to the improvement of the world around us. The thinking of the Gnostics disseminated widely throughout the Mediterranean and, with this drastic prejudice against the material reality, influenced many subsequent esoteric currents, from hermetic thought to the Qabbalah.

In contrast, as far as official Christianity is concerned, with the passing of the centuries the most active and proactive soul began to prevail, also thanks to the authoritativeness of charismatic figures such as Paul and Augustine. However, the spread of this thinking, from which the DNA of the technological West would have been born, first had to come to terms with the cultural and political superpower of the time: The Roman Empire.

References

1. Heraclitus, Diels-Kranz fragment, 22 B 91.
2. Quoted in Leucippus, fragment 2.
3. Aristotle, *Metaphisycs,* 1074 b 15 – 1075a 10.
4. *Ibid*, 1072 b 9–30.
5. Long A.A., *Hellenistic Philosophy Stoics, Epicureans, Sceptics*, Bristol Classical Press, Bristol 1974 (chapter "The Stoic Philosophy of Nature).
6. Stark R., *How the West Won. The Neglected Story of the Triumph of Modernity*, Intercollegiate Studies Institute, Wilmington, Delaware 2014.
7. Gospel of John 1, 1-3. All biblical quotes refer to the New King James Version.
8. Book of Genesis 3, 8.
9. *Ibid* 1, 28.
10. Reference 8, 2, 15.
11. Reference 8, 19.
12. Reference 8, 4, 20–22.
13. Book of Exodus 31, 2–5.
14. Gospel of Luke 13, 1–5.
15. Gospel of John 9, 3.
16. Book of Job 38, 4–5 and 37–38.
17. Book of Job 40, 4–5.
18. Ecclesiastes 1, 18.
19. *Ibid*, 16–17.
20. Ecclesiastes 12, 13–14.
21. Reference 18, 4–10.
22. Quoted in Stark, Reference 6.
23. Smith M., *Palestinian Judaism in the first century*, in Davis M., *Israel: its role in civilization*, Harper and Brothers, New York 1956, p. 71
24. Book of Sirach 16, 24–27. Translation from the United States Conference of Catholic Bishops.
25. Book of Wisdom 7, 15–22. Translation from the Revised Standard Version (RSV).
26. Mazzinghi L., *Il libro della Sapienza: elementi culturali* in "Ricerche storico-bibliche", 10, 1998, p. 186.
27. Tanzella Nitti G., *Teologia fondamentale in contesto scientifico*, Città Nuova, Rome 2018, p. 530.
28. Plato, *Timaeus*, 33c e segg.
29. Plato, *The Republic*, book X.

30. Reference 18 10–11.
31. Book of Genesis 6, 1–4.
32. D. F. Noble, *The religion of technology: the divinity of man and the spirit of invention*, Penguin Books, New York and London 1997.
33. *Ibid*, p. 10
34. Book of Daniel 7, 13.
35. Mark 10, 25; Matthew 19, 24; Luke 18, 25.
36. Matthew 4, 4; Luke 4, 4.
37. Luke 14, 26 (see also Matthew 10, 37).
38. John 12, 6.
39. Matthew 22, 21, Mark 12, 17, Luke 20, 25.
40. Augustine, *The city of God*, XXII, 24
41. Second Epistle to the Thessalonians 2, 1–3.

5

The Roman world and the "broken history"

5.1 THE 'PILLARS OF HERACLES' OF ANCIENT TECHNOLOGY

"The whole world was turned into a delightful garden by you, Romans [...]. Before your rule, life had to be hard, wild, like even now in the mountains [...]. Everywhere there are gymnasiums, fountains, temples, factories, schools [...]. All that is produced by the seasons and what the various regions, rivers, lakes and techniques of the Greeks and barbarians produce arrives here [in Rome] from every land and from every sea; if one wants to observe all these things, one must either go to see them travelling all over the world, or come to this city. In fact, it is not possible that what is born and is produced by each people is not found here even in abundance. There are so many cargo ships arriving here, carrying all the products from all the places, in every season, in every year, that Rome seems to be the general laboratory of the earth. And you can see so many shiploads from India and even from Arabia Felix, to assume that by now in these countries the trees have been left bare, and that they too must come here to look for their own products, in case they need something" [1].

This famous description of the Roman Empire at the height of its splendor was crafted by the Greek orator Elio Aristide (117–180) in the year 143, when the emperor was Antoninus Pius[1]. The rule of Rome extended without interruption over about six million square kilometers and included all the coasts of the Mediterranean Sea. It was undoubtedly a unique historical moment because, for the first time, this vast geographical area knew centuries of peace, economic prosperity and linguistic homogeneity.

---

[1] It is likely the speech was orated in the presence of the emperor himself.

It does not therefore seem out of place to talk about globalization. Thanks to the imperial political unity, goods, raw materials and people, as well as manuscripts and ideas, circulated throughout the Mediterranean with a speed and ease that until then had been unthinkable. From a technological point of view, the wealth of discoveries and innovations from Alexandrian Egypt were finally within reach of all humankind, at least for the understanding of the ancient West. With this, it would be reasonable to expect that a Golden Age of technological innovation would have dawned, especially given the abundance of material resources, but this did not materialize.

Despite the renewed political and commercial stability, as celebrated by Elio Aristide, the Roman Empire never succeeded in making the technological and cultural leap that could have led to the modern era. On the contrary, as Rodney Stark observed, the era could even be viewed as a *"setback"* in the *"ascent of the West"* from a technical and scientific point of view [2]. Provocatively, the American historian observes that, in terms of innovations, the only contributions provided by the Roman Empire to the birth of the contemporary world were the invention of cement and the spread of Christianity (the latter, however, was opposed by the imperial authorities) [3]. This is certainly a very negative, though not inaccurate, observation.

Even from an economic point of view, the shift from craftsmanship to modern industry, which a modern reader could expect owing to the abundance described in the speech by Elio Aristide, did not materialize. Thomas Pekáry noted that this is exactly as it had happened in ancient Greece, because *"very strict limitations were placed on the process of industrialization"* [4]. Yet one cannot truly question the essential theme of Aristide's oratory, despite its celebratory emphasis, which described what was in essence the first globalization in history.

How is it possible that, despite such a strong material basis, the journey from the ancient world to the modern one did not happen, even in an embryonic form? This question has left many authoritative scholars puzzled.

"Why was the urban civilization of Greece and Italy incapable of creating the conditions that could have ensured a continuous and uninterrupted journey forward from the ancient world to modern civilization?" [5]. Or, *"Why has there been no linear progress from the time of Hadrian to the twentieth century?"* [6]. These questions were posed by historians Rostovtzev and Walbank, respectively. More recently, summarizing the question, Aldo Schiavone spoke of a *"broken history"*, observing how effectively *"the Italian and European recovery, which will mark the true beginning of the new West civilization, will not look like the Roman imperial civilization in anything: the same humanists, who at the end of the fifteenth century theorized and realized the figurative rebirth of the ancient world in the Italian centers, now lived in an environment which did not look like the Roman environment, and they themselves were well aware of the distance*

that separated them from that lovingly evoked world" [7]. In short, Western civilization is in fact completely different to how it could have been, born in the Middle Ages from the ashes of the Roman Empire rather than from its evolution.

We cannot sidestep the failure of the ancient world to start a continuous technological development lasting up to the present merely by observing how the Roman Empire entered into a fatal crisis from the third century, which has shifted our attention to the causes of its decline. Even in the previous centuries, there were no relevant technological innovations and no economic models that had anything in common with industrial capitalism. What is striking is not so much that they failed to create a technological industry, but that they made no attempt to do so. Not only was there a lack of new scientific discoveries, but even the idea of using the existing Hellenistic machines to attempt to automate production was never considered. No attempt at using the profits to expand the size of craft shops and turn them into industries was even conceived.

Yet from a modern perspective, the abundance of raw materials, the availability of technological tools and the favorable political climate should have taken this evolutionary step just a little further than a simple automatism.

Analysis of the Roman Empire in the period of its maximum development reveals all the structural limitations of ancient civilizations. At that time, just like the mythical pillars of Heracles, a sort of ceiling of technological progress had clearly been reached. It was impossible to go beyond this within the limitations of the cultural and economic model of the time.

The broken history that Schiavone speaks about, namely the collapse of the Roman empire and the reconstruction of the West on a completely new basis, thus becomes a compulsory passage for the achievements of the modern age. Even when trying to imagine the continuation of the Roman empire beyond its historical limits, by eliminating with a single stroke the contingent causes of its crisis, it seems impossible to perceive the advent of industry and technology along such a path, even after centuries.

The reasons behind the existence of these 'Pillars of Heracles' of ancient technology, and the consequent long stagnation that characterizes the Roman world, have nothing to do with the lack of resources. Once again, the answer must be sought in the philosophical mindset of the time.

5.2 MATTER, SLAVES AND MACHINES: A BLIND ALLEY

The contempt that Roman intellectuals had for all craftsmen and, more generally, for anyone who dealt with activities that had to do with the world of dexterity and matter, was even greater than that of the Greek intellectuals.

"*All craftsmen practice a low profession: there is not even a shadow of nobility in a shop*" **[8]**. This harsh statement was voiced by Cicero, who openly expressed on several occasions the contempt with which the aristocrats of the time viewed manual work. Even the famous invective "*O tempora, o mores*" ("What a time! What costumes!") that he addressed against the corrupt governor Verre was not motivated by the outrageous abuse of power by the politician, as is commonly believed, but by Verre's habit of personally following the work of his goldsmiths' workshops dressed as a 'vile' craftsman "*with a dark tunic and a Greek cloak*" **[9]**. This was, apparently, a conduct and mode of dress unworthy of a respectable Roman politician!

In contrast to the modern, capitalistic point of view, the wealth that came from a craftsman's efforts was always considered 'dirty' money: "*The profits of all those mercenaries who sell not the work of their mind, but the work of their hand: the income coming from them is, in itself, the price of their servitude*" **[10]**. The basic premise, taken from Greek philosophy and brought to its extreme consequences, was that, in one way or another, anyone who worked with matter was in fact a slave.

Only occupations linked to literature, oratory and the abstract sciences were considered worthy of a free man, so it was not by chance that they were called the "liberal arts". To clarify this principle, Seneca was even harder than Cicero in his contempt for the world of manual work **[11]**. In his view, craftsmen "*have nothing to do with the true qualities of man*" **[12]**.

The ideal model of life for Roman intellectuals was *otium*, in which their time was filled with good reading, philosophical meditations, a taste for art and learned convivial disquisitions. The opposite was the denial of this kind of life, *negotium*, which included not only commerce, but also all crafts and practical activities.

This 'unworthy' world was bound to remain unrecorded by Roman historians and, unlike what happened in Athens, there were no exceptions even for those who dealt with prestigious works of art. In Rome, as Morel observes, "*the artist, or even the technician or the architect, are barely different from the craftsman to the eyes of those who influenced public opinion. With a few exceptions, they are not worthy to be remembered*" **[13]**.

For this reason, even Archimedes, the inventor par excellence according to Plutarch "*did not want to leave anything written on the construction of those machines, which even earned him a name and the fame of an almost superhuman and divine understanding. Convinced that the activity of those who build machines, as well as of those who perform any other activity aimed at an immediate utility, was both ignoble and vile, he devoted his most ambitious care only to studies whose beauty and abstraction were not contaminated by material needs... Although his inventions were varied and admirable, he asked his friends and*

5.2 Matter, slaves and machines: A blind alley

relatives that on his grave only a cylinder with a sphere inside was drawn. His epitaph was to be only the proportion between the volume of the containing solid and that of the contained one" **[14]**.

This view of the world had a devastating impact on technological innovation. The conflict between intellect and matter caused a deep separation between abstract science and any attempt to apply it in practice.

The only dissenting voice was that of Posidonius (135 BC − 50 BC) who, perhaps due to his Greek origins, tried to bridge this gap by tracing the origin of all material inventions back to the rationalist contribution of the ancient philosophers: "*All these things were invented by the scholars, who however entrusted them to lower executors, considering them too low to deal with them personally*" **[15]**. While he also considered technological innovation to be unworthy of the direct attention of the wise, Posidonius recognized the existence of a relationship between abstract theory and applied science, thus ennobling the latter. His stance, however, was met with sarcasm from other Roman intellectuals, who could not conceive of any possible connection between the two worlds. As Seneca ironically noted, alluding to Posidonius, "*He almost said that even the craft of cobblers was invented by the wise*" **[16]**. Yet this dualistic standpoint is not sufficient in itself to explain the mystery of Roman stagnation, as discussed in the previous section. We must add a last key piece, of an economic nature, to this picture: the role of slavery.

In the Roman civilization, and because of the first Mediterranean globalization, slavery had assumed a completely new nature, becoming a structural and pervasive element of the economy. Used more than anything else to increase production volumes of the individual shops and in the huge public building projects, slavery suddenly became an essential pre-condition, without which the entire productive world could not operate autonomously. Schiavone defines the Roman slave system as "*the actual propulsive center of the Mediterranean economy, without alternatives, neither theoretical nor practical*" **[17]**. Moreover, at the time of Augustus the market price of a slave was particularly low (the equivalent of 3−5 square kilometers of land to be used as vineyard **[18]**) while, because of the many victorious wars, the available supply was extremely high. In the Italian peninsula alone there were 2−3 million slaves, equivalent to about 35 percent of the total population. Every productive activity of the Roman state was based on this forced labor market, to the point that quite soon, in the collective imagination, slavery became part of the natural order of the universe.

Even the intellectuals, who were particularly concerned about respecting human dignity and about the principles of social justice, unanimously accepted the presence of slaves, without whom the existence of civilization itself would have been impossible. Regardless of its size, every attempt at rebellion was promptly suppressed with ruthless brutality because the preservation of social order was at stake. Even in the case of a domestic and isolated revolt, such as the murder of prefect Lucius Pedanius Secondus by a slave, the senate was inflexible in enforcing the law, in this case to the point of sentencing all the other 400 servants to capital punishment, even though they were manifestly innocent[2].

Nowhere among the entire literature of the time was there a voice against the practice of slavery, because no author could imagine an alternative that would also allow the vast Mediterranean empire to survive.

Not even Posidonius, who had limited himself in the Republican period to criticizing certain particularly cruel and degrading harassments of slaves[3], was an exception to this. Posidonus had never hoped for the end of slavery, but wished only that this practice was both more humane and of greater social utility. The alternative, utopian and unfeasible, was to return to the archaic and patriarchal society, based on economic self-sufficiency.

In fact, the Romans never conceived any practical alternative to slavery. This type of workforce was extremely convenient, and was inextricably bound to discouraging technology, machines and automation. The margins for a radical change of perspective were minimal, given that this economic advantage was accompanied by the deep cultural disregard for materiality and craftsmanship. This was a fatal short circuit, which led to many centuries of stagnation of the technological evolution.

5.3 THE STRUCTURAL LIMITATION OF THE ROMAN ECONOMY

Essentially, technology could not develop in the Roman Empire because of the pervasive use of slavery. In considering this, however, it could reasonably be argued that at the time of the American Civil War, slavery and technological evolution coexisted in the economy of the Southern Confederation in a very efficient production model. The difference is that beyond the division of tasks between man and machine, both human and mechanical factors are essential in a modern production process.

Why did this not happen in ancient Rome? Beyond the aristocratic disregard for matter and the competition of slave labor, the first real cause of the technological underdevelopment of the Roman Mediterranean can be found in the failure to develop a capitalistic model.

---

[2] The episode happened in the year 61 AD and is very well described in Tacitus, *Annals* 14, 42–44.

[3] The ideas of Posidonius are indirectly preserved in the *Bibliotheca Historica* of Diodorus Siculus.

As already noted above, this was not an economic problem in the strictest sense. The abundance of resources alone, as described by Elio Aristide, was a more than sufficient argument to demonstrate how the conditions for the birth of capitalism had actually been present for centuries in the Roman world. The problem was instead more of a social and psychological one. Even the entrepreneurs, merchants and craftsmen, so despised by Seneca and Cicero, never worked towards reinvestment in, or the growth of, the company, but merely a lifetime income, a notion typical of the agrarian nobility. For these social categories, the status symbol of reference always remained that of the landowner.

This attitude was the true cause of the non-transition from craftsmanship to industry and, on a larger scale, of the structural limitation of the entire Roman economy. All the income generated by the companies was systematically removed from the economic cycle, without any form of reinvestment.

With the expansion of the empire and the spread of well-being, this logic generated more and more evident wastage and led to the systematic closure of many potential production units, laying the foundations for the collapse of Roman power. Taking advantage of the slave labor force, the affluent sections of the population reached a point of saturation beyond which they no longer had any interest in expanding the size of their business.

As a consequence, starting from the third century after Christ, an ever-greater portion of the population of the city of Rome and the surrounding districts simply stopped doing any kind of work. As Schiavone observes, based on the argument that the work of slaves guaranteed an income, "*it was cheaper to feed these men and their families in a completely parasitic way, than to use their work in any productive function*" **[17]**. In return for their food, this vast unproductive class sold its political support to the highest bidder.

As Mihail Rostovtzev effectively explains, "*every new city was transformed into something like a new hive of drones*" **[19]**. As a result, the peripheral provinces had to work almost exclusively to maintain these unproductive realities, and were deprived of resources they could have used for economic initiatives **[20]**.

As is evident, this approach is completely opposite to that governing the modern capitalist economy. In a system which can afford not to employ its biggest potential workforce, and which has no interest in the actual improvement of production, technological innovation is clearly of secondary importance.

5.4 NOT JUST SHADOWS: IMPERIAL TECHNOLOGY

Thanks to the Roman Empire, the evolution of Western technology suffered a serious setback from which it would only recover in the Middle Ages. Compelled to traverse the "broken history", the entire European civilization had to start afresh from new bases.

However, the five hundred years of history between the ascent of Augustus (31 BC) and the beginning of the Middle Ages (476 AD) were also the period in which many Alexandrian technologies found an effective application and dissemination. Undoubtedly, they were grossly underused, because the Romans were little interested in investigating the scientific laws underlying such tools, but merely that they worked in the desired way.

Nevertheless, Roman civilization was undoubtedly the most technological of the ancient world. This was a 'silent' phenomenon and one that was openly rejected by the intellectual world of the time, but at the same time, one that was widespread. To understand this schizophrenia between rhetoric and reality, we must abandon literary sources and look to the world of archaeology.

As Jean-Paul Morel explains, *"every observer of the Roman world, provided he is consistent and in good faith, cannot help but be impressed by the enormous quantity of artefacts that were produced, exchanged, consumed. Archaeology tirelessly puts in evidence the surviving specimens, just the tip of the iceberg. In short, the archaeologist, the historian, or the simple amateur must recognize that the craftsmen, with their products and services, play a role in the Roman world that they could not have suspected just reading the ancient authors"* [21].

Technology, like slavery, was essential for everyday life. The administration of imperial cities necessarily required the contribution of a crowd of technicians, engineers and builders, not just slave labor.

These professions were hierarchically organized, and masters and apprentices worked side-by-side in the shops, trying to keep their working techniques secret and handing them down only orally. The gap between the craftsmen and the 'technologists', i.e. the intellectuals such as Vitruvius, Pliny the Elder or Frontinus who spent their time writing about these subjects, was enormous. These writers very often knew this world only from the outside, without actually mastering the details of what they described in their books. So, to understand Roman technology, one must directly study the remains of Roman artefacts.

The only field in which high ideals and practical work came together was military activity. All the soldiers of the Roman army, from the commander-in-chief to the last infantry soldier, were required to master the technological arts. The Roman army was perhaps the first to have a corps of engineers permanently organized in each legion. This unit of specialists was formed by engineers, architects, surveyors, carpenters and locksmiths in permanent service and was under the command of an officer, the *praefectus fabrum*. The most famous of the Roman military engineers was Apollodorus, who worked at the time of Emperor Trajan.

In the following sections, an overview of Roman technology will be traced, trying to describe this world, often forgotten by the literature of the time, in some detail.

5.5 METALLURGY

The age of iron undoubtedly predates the Roman period and the use of this metal continued well beyond the areas in which the Roman civilization flourished. However, the Roman civilization was the first to make extensive use of metals, mainly of iron, a use which expanded until the fall of the Roman empire.

Mining activity was quite intense, particularly in the Imperial Age, and attracted much criticism from the same intellectuals who regarded all material activities with contempt. There is no doubt that large quantities of timber were needed to melt metals, which, combined with its use domestically, for shipbuilding and for construction in general, put a severe strain on the sources of supply. The problem had been present in the Greek era, or even before, and was inherited by the Romans. Plato had written extremely critical pages on the deforestation of the Athens region, on the resulting erosion of the upper layer of soil and on the consequent loss of soil fertility. In his *Crizia,* the mountains of Attica were described as *"bare bones almost like those of a sick body"* [22]. However, the environmental damage of mining was not limited to deforestation, which also had other causes, but included the excavation of large areas that caused the collapse of entire mountains. To use Pliny's words, *"The victorious miners observe the collapse of nature"*.

The pollution of the air, water and earth caused by mining activities was severe, in particular as far as pollution by heavy metals such as lead or mercury, arsenic and antimony, was concerned. According to Pliny the Elder, but also Cicero and Seneca, behind everything there was always greed: *"the things that ruin us and lead us to the underworld are those that the Earth has hidden and concealed in its bosom, things that are not generated in a moment; so our mind, projecting itself into the future, considers when we will, over the centuries, exhaust them, and how far our greed can penetrate"* [23, 24].

These considerations, which echo present-day alarm concerning the depletion of raw materials, have significant moralistic overtones and are based on a fundamental misunderstanding of the economic aspects. To attribute all such problems solely to human greed is, without doubt, reductive and misleading.

Actually, the use of metals is a phenomenon that has affected not only the likes of military technologies, but all applications in general and agriculture in particular. The widespread use of the metal plough (which is an older technology, but became more common only thanks to the increased scale of metal production in the Roman period), and the moldboard plough which became common in the Middle Ages, enabled the cultivation of lands which could not previously have been worked, or at least greatly increased agricultural productivity. This issue also highlights the ambivalence of the Roman intellectuals towards technology. The same Pliny who spoke against the greed that pushes humankind to extract metals from the earth also left detailed descriptions of the various types of plough, illustrating in detail the advantages of innovations in the agricultural field, including those related to the use of metal.

However, despite the increased scale of metal production, its use remained quite limited. In Roman, and then Medieval and Renaissance times, machines were built mainly of wood. It took until the modern era to see machines built of metal, particularly iron (cast iron and steel). In general, this change related mainly to the construction of steam engines and occurred in the nineteenth century. For the construction of ships, the widespread use of metals occurred still later (late nineteenth century) and for aeronautics it was delayed until well into the twentieth century.

5.6 MILITARY TECHNOLOGY

As noted, the Roman army included a structure dedicated to the construction and maintenance of a number of war machines. Even in this field, the Romans invented little, but they realized pre-existing devices and machines on a scale never seen before. A typical example is that of siege machines, which had been used extensively by the Greek and the Eastern empires and then in that synthesis of East and West which was the army of Alexander the Great. However, they were used by the Romans on a much larger scale. Vitruvius described mobile towers, equipped with drawbridges higher than the highest walls, weighing more than 100 tons (Fig. 5.1).

The wooden parts of the various machines were prepared on site to be added to the metal parts, which were stored after use. The artillery used twisted animal strings or bundles, or metal or wooden bendable springs, to store the energy that was then imparted to the projectile. The most advanced of these was the *ballista*, which dates back to the Hellenistic period and is said to have been invented in 399 BC by a group of 'scientists' working for Dionisius the Elder of Syracuse in developing a new weapon for use in the war with Carthage. Even in this case, however, the Romans would have done nothing more than utilize and slowly refine a pre-existing technology.

The Romans positioned ballistas on wagons so that they could be moved easily and, according to some descriptions by Tacitus, not only used them as siege machines, but also as field artillery during battles: "*A huge ballista, belonging to the XV legion, began to cause serious damage along the front of the Flavians by throwing large stones, and would have caused extensive destruction if not for the splendid courage of two soldiers, who, taking some shields from the dead soldiers and then disguising themselves, cut the ropes and springs of the machine*" **[25]**.

Metal armor, mainly used by cavalry, consisted of plates connected to each other in such a way as to allow a certain freedom of movement to the rider and the

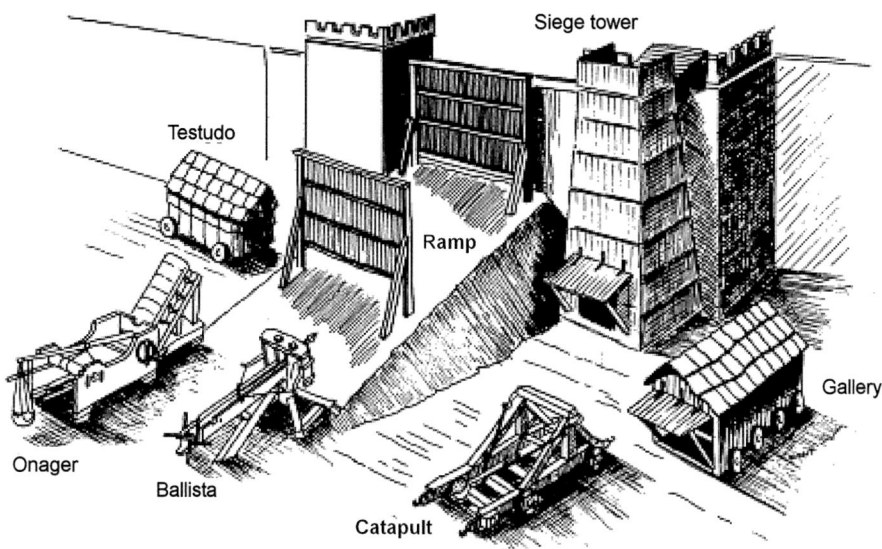

Figure 5.1. Some siege machines and structures used by the Roman army. [Image from https://upload.wikimedia.org/wikipedia/commons/d/d7/Roman_siege_machines.gif.]

horse. They were introduced at the time of Alexander the Great and were then used by various Eastern armies, such as the Parthians, and finally by the Romans. The use of heavy armor by cavalry also required the introduction of devices such as the saddle and stirrups, which would give stability to the rider, and of horseshoes, without which the animal was unable to carry the weight of an armored knight easily. Until very late, however, the Roman army mostly relied on its infantry, which was less heavily armored.

5.7 CONSTRUCTION INDUSTRY

One of the areas of technology in which Romans were at their best was engineering buildings, certainly as far as public buildings, transportation systems (roads, bridges, tunnels, aqueducts) and military architecture are concerned. The round arch is typical of Roman engineering. It was a much older development (there are examples dating back from the third or even the fourth millennium BC), but only became widespread in Roman times and then continued to be very common in the Middle Ages with Romanesque architecture. The arch itself was not a Roman invention, but Romans built a huge number of arches, using them for various applications and realizing them on a scale, and a size, never seen before.

88 The Roman world and the "broken history"

Certainly, the knowledge of hydraulics which had begun to accumulate in Hellenistic times made it possible to create pressurized aqueducts across valleys, without having to build the imposing arches typical of Roman aqueducts. However, where possible, the Romans preferred the perhaps more expensive but conceptually simpler solution of transporting water in canals with a constant and very slight slope (Fig 5.2a).

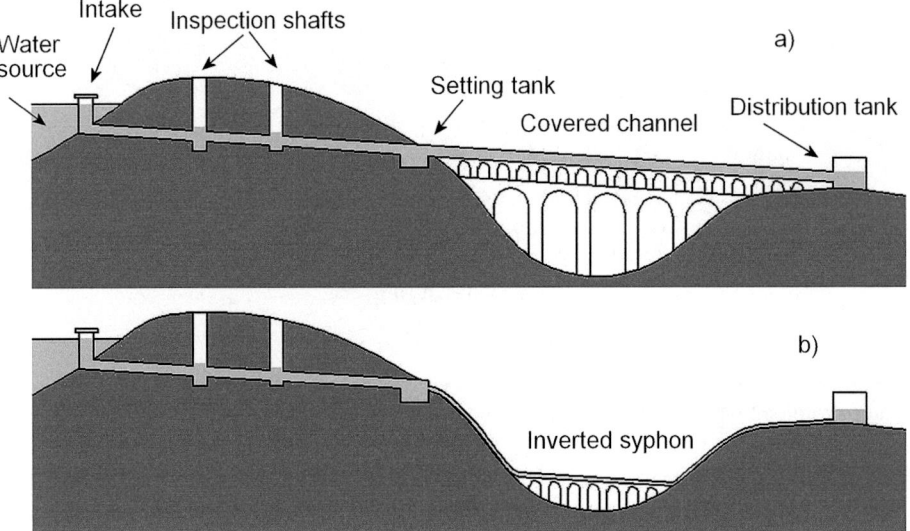

Figure 5.2. a) Sketch of a typical Roman aqueduct. b) The inverted siphon, which required pipes able to withstand higher pressures, was seldom used in Roman times.

The inverted siphon (Fig. 5.2b), which made use of smaller structures than conventional aqueducts, was used by the Romans only when it could not be avoided to keep water leaks to acceptable levels, mainly due to the frequent and costly maintenance required by the pressurized pipes.

To keep the canal almost horizontal, it was often necessary to build very high aqueduct arches, a famous example being the aqueduct of Segovia (Fig. 5.3) built in the second half of the first century AD, which at its highest point reaches 28.5 meters, including about six meters of foundations. Connected to the aqueducts, there were many dams and cisterns built by the Romans throughout the territory they controlled.

Another area in which the Romans created unique public works is the construction of roads. Thousands of kilometers of paved roads (it is estimated that the maximum extension of the Roman roads reached 85,000 km) covered all the

Figure 5.3. The aqueduct of Segovia. [Image from https://upload.wikimedia.org/wikipedia/commons/f/f4/Aqueduct_of_Segovia_08.jpg.]

territory that belonged to the empire and often included many bridges and even some tunnels. Roman roads were so well built that many remained in use for a millennium after the organization that took care of them had dissolved with the fall of the empire. Many argue that roads of comparable quality to those constructed by the Romans did not appear again until the nineteenth century. It should be noted, however, that Roman roads were designed mainly for pedestrian traffic, although they were also travelled by men on horseback and occasionally by carts.

All the cities sported many public buildings, including the religious (temples) but above all civilian (theatres, amphitheaters, covered markets, basilicas, baths, etc.). The size of some of these buildings was truly impressive.

Even the houses were remarkable, in some cases for the luxury and comfort granted to their inhabitants. The introduction of glass windows helped significantly with maintaining a pleasant temperature inside the building, while at the same time allowing the light to enter. Inevitably, the high cost of glass windows limited their use to the homes of the very rich.

Most of the people, particularly in Rome and in some large cities, lived in *insulae*. These were buildings that, although impressive for their size and having several floors, were built economically with cheap materials and often disregarding the rules of building safely, which were well known even then. Collapses and fires were frequent, and where they did not endanger the lives of their inhabitants, the buildings were still very uncomfortable and unhealthy. Pliny and Seneca's invectives against human greed were certainly applicable to the behavior of owners and contractors.

Undoubtedly, the construction of architecture of the size of aqueducts, basilicas and even the most common *insulae* required the use of construction machinery, such as cranes, lifting systems, and machines to drive foundation piles, as well as instruments such as levels, the *archipendulum*, the *groma*, and more complex optical instruments like the *dioptro* invented by Heron. The, cranes, driven by treadmills (see below), were normally able to lift blocks of stone or other objects of up to several tons.

5.8 ENERGY PRODUCTION

It is difficult for us to realize the scarcity of energy with which the world lived until very recently, and the Roman world is no exception. Most of the energy came from the muscles of animals and humans. Given the widespread use of slaves, those who supplied energy to the machines of the Roman world were often human beings and the choice of whether to use humans or animals often depended on a mere economic calculation: how much the purchase of a slave would cost, how long he would last and how much he would eat, in comparison to an animal, to perform a certain job. This situation lasted for the entire duration of the republic and up to the latter stages of the empire, when the reduced military capability of the army led to a lack of slaves and Christianity began to change the collective attitude towards slavery.

The most common machine used to transform muscle energy into mechanical energy was the treadmill (Fig. 5.4a, which shows a treadmill used in the late Renaissance to operate a winch), which has remained in use for nearly two millennia, even if it is now used only in small animal cages, for example for hamsters.

Other devices were less sophisticated (such as the one shown in Figure 5.4b, also taken from a much later drawing, but certainly used in Roman times). When the mechanism was operated by a human instead of an animal, it was possible to use a crank or a pedal, which were also used in Roman times.

Gears or a cable transmission were often interposed between the device driven by the animal or human being and the machine that used the power (a mill, a

5.8 Energy production 91

Figure 5.4. a): A treadmill driving a winch, designed by Antonio Ramelli (in *The different and artificial Machines*, 1588); b): Grindstone driven by a horse, drawing by Georg Andreas Boeckler (in *Theatrum Machinarum Novum,* 1661) **[26]**.

winch, a grinding wheel, etc.). The Roman gear wheels were generally made of wood, such as the pin wheel and the lantern wheel shown in Figure 5.5, and their efficiency was rather low. However, even gear wheels were not a Roman invention, since they were already in use in the Greek era. Small wheels, transmitting very limited power, had been used in instruments and were made of metal. As demonstrated by the 'Antikythera mechanism' found in 1902, planetary gears were also used. The Antikythera mechanism can be traced back to the beginning of the first century BC and is clearly a product of Hellenistic technology. For a detailed description, see Lucio Russo, *The Forgotten Revolution*, Springer, Berlin 2004.

The only forms of mechanical energy used in antiquity were hydraulic energy and wind power. Given that the main use for such devices was often to grind wheat, we tend to think of water mills and windmills, but their use was much more general and ranged from driving grinding wheels, saws and other cutting tools, to the actuation of bellows, shredders, mortars, winches, etc., or simply to raise water.

In some cases, as for operating saws, it was necessary to transform the rotary motion of the water wheel, or of the windmill, into a reciprocating motion, which required the use of a connecting rod-crank mechanism (Fig. 5.5).

92 The Roman world and the "broken history"

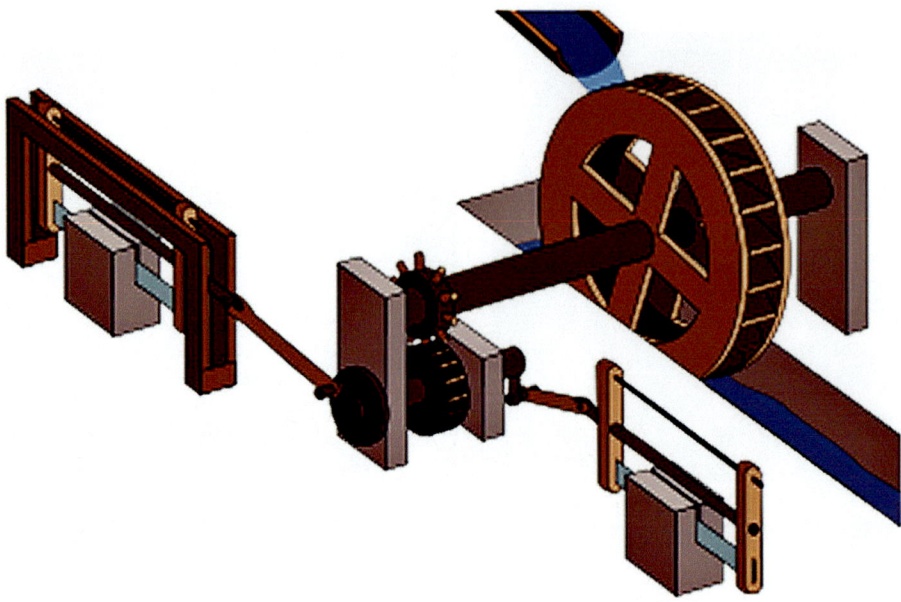

Figure 5.5. Scheme of the 'mill of Hierapolis', in Asia Minor (3rd century AD), used to drive two saws to cut stone blocks. It is believed that this is the oldest machine to use crank-connecting rod mechanisms to transform a rotary motion into reciprocating motion. Note also the pin wheel and the lantern wheel. [Image from https://upload.wikimedia.org/wikipedia/commons/b/b3/R%C3%B6mische_S%C3%A4gem%C3%BChle.svg.]

Both the water wheel and the windmill probably originated in the regions to the east of the Roman Empire, before Rome extended its rule over them, so once again these cannot be considered as Roman inventions. However, their widespread use took place over time in the Roman environment. There has been much discussion about how effectively the great spread of slavery slowed down the use of these forms of mechanical energy, and many argue that the water and windmills only became common in the late empire mainly due to the spread of Christianity.

On one point there are few doubts. The ancient world never saw the creation of actual thermal machines, in which the heat from a fire was transformed into mechanical energy. The only example, the aeolipile, described by Heron probably in the first century AD, is not so much a precursor of the action turbine, but what we would today call a technological toy, given the impossibility of obtaining sufficient power to make it feasibly practical. Similarly, other devices described by Heron himself, such as a system to open the doors of a temple automatically, had no practical developments. The realization of thermal machines would have required further development of the scientific technology that had begun in the

Hellenistic environment but, as already mentioned, was aborted with the expulsion of the Greeks from Alexandria.

Fuels, such as wood and (seldom) coal, were then used directly to produce heat, for heating homes and in particular the water of the baths, for cooking and for those industrial processes that required heat (baking bricks, ceramics, metallurgy, etc.), but not for the production of mechanical energy. However, the scarcity of wood in the vicinity of the cities and the high cost of transportation were constant in the ancient world and worsened in the late Roman period.

5.9 AGRICULTURAL TECHNOLOGIES

As well as being an economic activity, agriculture was also considered a lifestyle by the Roman intellectual elite, one that was deemed far higher than craftsmanship and commerce. In his treatise *On Duties*, Cicero wrote *"Among the occupations in which a revenue is assured, none is better than agriculture, neither more profitable, nor more pleasant, nor better suited to free men."* One cannot fail to note the contrast with the servile character attributed to craftmanship by these same intellectuals.

As already mentioned, one serious problem which hindered agricultural development throughout the ancient world was the lack of knowledge of the anatomy of the animals used to draw carts and agricultural machines, particularly the horses. As noted, horses (or, more frequently, donkeys) cannot be harnessed in the same way as oxen, which push on the yoke with their shoulders, so a strap passing over the animal's chest was used. This pressed against the horse's windpipe and choked it, just when the animal had to exert the maximum effort. This problem, which was solved only towards the end of the Middle Ages, meant that the power available for all operations requiring animal traction, (ploughing or drawing carriages, but also in general to supply power through devices such as the grindstone in Fig. 5.4b), was very limited. The treadmill (Fig. 5.4a), in contrast, was not affected by this problem.

It has already been said that the introduction of some improvements to the plough made it possible to cultivate land which previously could not be exploited, particularly in northern Europe. One of the few technological inventions of Roman times seems to be the combine harvester, a machine capable of both reaping the wheat and threshing it, which makes it possible to speak of a real mechanization of agriculture in Roman times (Fig. 5.6). However, it is not certain whether this invention can actually be ascribed to the Romans, or whether it was introduced by the Celts. Roman farms were also equipped with presses for the olives, but once again these presses pre-date Roman times and became widespread only in the Roman environment.

94 The Roman world and the "broken history"

Figure 5.6. Frieze depicting a Roman combine harvester. [Image from https://upload.wikimedia.org/wikipedia/commons/3/36/Roman_harvester%2C_Trier.jpg.]

Even in agriculture, like in many other sectors, the most important Roman innovations were not so much technological, but organizational and managerial.

5.10 MEANS OF TRANSPORTATION

Throughout the ancient world, land transportation of heavy and bulky goods was always difficult, so water transportation (by sea or rivers) was used wherever possible. Things did not change substantially in Roman times, although the construction and maintenance of an extensive road network may have made land transportation easier. However, the incorrect method of harnessing horses greatly reduced the traction available to carts and wagons. A typical Roman light cart was the two-wheeled *birota*, with a payload of a hundred kilos, while the *plaustrum* was a heavy wagon, sometimes four-wheeled, capable of carrying heavier loads. Interestingly, the two-axle wagons were equipped with a steering front axle, although it is not known when this essential innovation was introduced.

Regardless, the speed at which you could travel by road was quite low, not more than ten kilometers per hour even in the most favorable terrain. The average distance travelled daily by imperial couriers was 70–75 km, although longer distances could be traveled in exceptional cases, such as in 48 BC when Caesar reached the Rhone from Rome by covering 150 km a day.

Transportation by sea or by river could be easier and faster, particularly where heavy loads were concerned. A ship could travel at an average speed of 3–5 knots, but it is difficult to think of travelling by sea more than 100–200 km per day. In addition, the variability of winds and weather conditions meant that the percentage of time that it was actually possible to travel was not high.

The Romans were initially well aware of their inferiority in naval warfare compared to many of their neighbors and adversaries, particularly the Carthaginians who had inherited the experience of the Phoenicians, and the Greeks. To overcome this inferiority, they launched large and expensive naval construction programs and introduced the *rostrum* (of Etruscan invention, according to Pliny the Elder) and the *corvus*, a boardwalk that, as its creators intended, enabled a sea battle to be transformed into a clash between infantries. In fact, the invention of the *corvus*, which this time was certainly due to the Romans, helped them to victory in the battle of Milazzo against the Carthaginians and later made the Romans masters of the *Mare Nostrum*, the Mediterranean Sea.

In imperial times, Romans started building huge ships, even if the largest, like the so-called 'ships of Nemi' could be considered more ceremonial or as means of propaganda to exhibit the power of the empire than as true ships to be used for war or peace.

The Roman ships of the imperial period were equipped with catapults, ballistae and other artillery. For the time, they had very advanced refinements, such as ball or roller bearings used in winches, although we do not know how usable these wooden bearings actually were. Although it had probably been used in earlier times in the Aegean Sea, the auric sail, a trapezoidal sail arranged aft of the mast, is often attributed to the Romans. The Romans also used the lateen sail (or latin rig), although this term is sometimes attributed to the use of the term *latin* in the sense of 'easy', or, more likely, to a corruption of '*alla trina*' because of its triangular shape, and therefore does not refer to its Roman origin.

The Roman warships were rowing ships, like the warships of all the ancient populations of the Mediterranean Sea, while sails were mainly used on cargo ships. With their ships, Romans ventured far beyond the Pillars of Heracles, sailing along the Atlantic coasts of Europe and Africa. Many say that Roman ships visited the American coasts and while it cannot be ruled out that some Roman ships, perhaps involuntarily, may have traveled to such distances, we have no evidence of transatlantic journeys by the Romans. What is certain is that if they did so, then these were occasional trips which left no lasting consequences in history. In order to be able to cross the ocean regularly and safely, much different ships would be needed to those used by the Romans, or those on which the Vikings ventured on their transatlantic journeys centuries after the end of the Roman period. Such ships only became available in the fifteenth century AD.

References

1. Aelius Aristides, *To Rome*.
2. Stark R., *How the West Won. The Neglected Story of the Triumph of Modernity*, Intercollegiate Studies Institute, Wilmington, Delaware 2014.
3. *Ibid*, p. 9
4. Pekàry T., *Die Wirtschaft der griechisch-römischen Antike*, Steiner, Wiesbaden 1979.
5. Rostovtzev M.I., *The Social and Economic History of the Hellenistic World,* Clarendon Press, Oxford 1941.

6. Walbank F.W., *The Decline of the Roman Empire in the West*, University of Toronto, Toronto 1969, p. 36.
7. Schiavone A., *The End of the Past: Ancient Rome and the Modern West*, Harvard University Press, Harvard 2000.
8. Cicero, *On Duties* I, 42, 150.
9. Cicero, *Contra Verrem*, II.
10. *Ibid.*, I, 42, 150.
11. Seneca, *Lettere a Lucilio* 80, 2.
12. *Ibid.*, 80, 20.
13. Morel J.P., *L'artigiano*, Laterza, Rome-Bari, Italy, 2013, p. 18.
14. Plutarch, *Life of Marcellus* 17.
15. Quoted in Seneca, *Moral Epistles to Lucilius* 90, 25
16. *Ibid*, 90, 23
17. Schiavone A., *The End of the Past: Ancient Rome and the Modern West*, Harvard University Press, Harvard 2000.
18. *Ibid*, p. 127.
19. Quoted in Dawson C., *The Making of Europe: An Introduction to the History of European Unity*, Sheed and Ward, London 1932, reissued by the Catholic University of America Press, Washington DC, 2003.
20. *Ibid*, pp. 85–86.
21. Reference 13, p. 8.
22. Weeber K.W., *Smog über Attika*, Artemis Verlag Zürich and München 1990.
23. Reference 8, 1, 42
24. Pliny the Elder, *Natural history*, 33, 1-3
25. Tacitus, Historiae, III, 23
26. Bassignana P., *La cultura delle macchine*, Allemandi Editore, Turin, Italy 1989.

6

The Middle Ages: "Dark ages" or the dawn of technology?

6.1 THE INVENTION OF THE DARK AGES

For almost six hundred years, "*barbarism, superstition and ignorance covered the face of the earth*" [1]. "*The inhabitants of this part of the earth (...) lived in a condition worse than ignorance*" [2]. "*It was the era of the triumph of barbarism and religion*" [3].

These are not isolated condemnations. Those who expressed these opinions on the European Middle Ages were three of the most authoritative intellectuals of the Enlightenment period, Voltaire, Jean-Jacques Rousseau and Edward Gibbon, who for centuries influenced the collective imagination of the Western world with their thinking.

Even the phrase, "Middle Ages", is an expression that suggests a provisional situation, devoid of an identity and defined only by its nature as an interlude between two phases of splendor, the Roman Empire and the Italian Renaissance. Bertrand Russell also focused on these alleged "*dark centuries*", arguing that "*it is not inappropriate to define them as such, particularly when compared with what came before and what came next*" [4]. More recently, the historian William Manchester has gone even further, talking about "*a world lit only by fire*" [5]. This is not because in Roman times and the Renaissance some form of electric light was widespread, but because he believed, as with his famous forerunners, that between 400 and 1000 AD the light of human rationality had almost completely disappeared in Europe.

Violence, war and scientific involution. This all paints a portrait akin to that of a post-apocalyptic world, so much so that between the years 1960–1980 there was talk of a 'future Middle Age' to describe the conditions in which we could have been forced to live if a nuclear war or another global catastrophe had destroyed our civilization. A world like that of *Mad Max*, to mention one of the most famous movie series of the post-apocalyptic genre.

It is not by chance that the Middle Ages is considered to have begun in the year 476, the date of the "end of the world", as the fall of the Western Roman Empire was described by some chroniclers of the time[1]. The armed invasion of the barbarians, the semi-primitive populations of Germanic origin, was regarded as the collapse of civilization. Trade disappeared, the cities lost their populations, superstition and the religious fanaticism of the Christian matrix came to the fore, the Greek scientific conquests were forgotten and illiteracy dominated.

However, this picture is as false as it is vivid. In reality, it is a clamorous historical fake, completely devoid of historical grounds. Just as in the case of the alleged antithesis between the Christian religion and rationalism, described in Chapter 4, this is a common misconception that is still being presented as a historical truth in many schools, although it is now widely refuted by the scientific community. For example, the authoritative Columbia Encyclopedia now rejects the use of the expression *"dark ages"* as historians *"no longer consider that the medieval civilization was so gloomy"* [6].

6.2 FOUR COMMON MYTHS ABOUT THE EARLY MIDDLE AGES (476–1000)

If ideological prejudices are set aside and the Middle Ages are observed from the historical point of view, very little remains of the Enlightenment myth of the 'Dark Ages'. In particular, there are four common misrepresentations that must be shown to be wrong before being able to ascertain the pivotal role of the Middle Ages in the history of the western technological world.

Illiteracy

First of all, it is incorrect to speak of a spread of illiteracy. Given that, as noted by William Harris, the number of people who could read and write did not exceed five percent even in the Imperial Age, the statement is actually the result of a macroscopic error of perspective [7]. The Early Middle Ages (476–1000) was an era of great linguistic transition. Written Latin, as we know it, dates back to the period between 100 BC and 150 AD, in which it was used by the great authors of the classical era.

---

[1] In his introduction to the *Commentary on Ezekiel*, Jerome says that *"the light of the world is extinguished: The Roman Empire has been beheaded"*, and in the *Epistle 127* he even talks about a supposed *"carnage"* while comparing the episode to the fall of Troy as described by Virgil. These are mere hyperboles, meant to stigmatize the strategies of the Roman-barbarian general Stilicho.

6.2 Four common myths about the early Middle Ages (476–1000)

Since that time, its grammar remained strictly crystallized. Over the centuries, the evolution of the oral language actually spoken in everyday life led to the physiological onset of increasingly notable differences, to the point of delineating new languages, the so-called 'vulgars'. The phenomenon was due to the everyday use of the spoken word, and has little or nothing to do with the so-called 'barbarian invasions'. Proof of this is evident in the fact that the various vulgars spoken in Italy, France and the Iberian Peninsula were, in all respects, 'Romance' languages of Latin derivation that were hardly influenced by the Germanic languages.

On the other hand, the decline in the number of people able to master written Latin was almost exclusively because it became further removed from the everyday spoken language. At the same time, however, there were still no rules for putting the vulgar into writing, a language that was considered unworthy to be used in official documents. This is an 'illiteracy' that cannot be considered as cultural backwardness: how many Britons currently understand the linguistic and grammatical rules of Chaucer's English? It is not easy to write in a language that is never used colloquially.

The transition from Latin to vulgar languages was particularly slow and there is evidence of several instances of bilingualism. For example, the oaths of Strasbourg (842) by the Carolingian sovereigns Charles the Bald and Louis the German were proclaimed in a sort of proto-French, so that they could also be understood by the troops; while the Placito Capuano (960–963) was a declaration made by some farmers in the current language of southern Italy at that time in the context of a judicial controversy[2].

The alleged wave of illiteracy was reversed only in the Late Middle Ages, when some particularly influential authors (Dante, Petrarca, Chaucer) chose to write their works in the new vulgar languages, thus affording them literary dignity.

One could argue that this is a partial explanation, because during the first centuries of the medieval period even the cultural elite, who had mastered Latin, actually stopped producing new literary works. In reality, the reasons behind this choice are many, but they have very little to do with an alleged regression of civil society:

- on a literary level, the cult of classical authors spread, and they were considered as qualitatively unattainable. Initially, it was therefore considered better to preserve the existing works by copying them, while limiting the new productions to the encyclopedic comment and marginal annotations. "*We are dwarves on the shoulders of giants*", as Bernard of Chartres (1070–1130) effectively summed up;

---

[2] For further information about the slow transition from written Latin to the 'Romance languages' and the bilingualism, see L. Renzi, A. Andreose, *Manuale di linguistica e filologia romanza*, Il Mulino, Bologna, Italy 2003.

- the meeting between heterogeneous cultures such as Christian (finally free from imperial repressions), classical Roman and Germanic, led to a phase of complete reconstruction of literature. Consequently, some areas, such as poetry, attracted less interest compared to others, such as theology;
- with the disintegration of the empire, the bureaucratic apparatus disappeared, and the structure of society changed radically to the detriment of the higher classes. Many people saw their income, linked to a position within the imperial bureaucracy or with the possession of large estates, either vanish or at least shrink due to the arrival of new populations. At the beginning of the Early Middle Ages, the figure of the professional intellectual, dedicated exclusively to *otium,* almost completely disappeared. In short, a part of the population did not stop dealing with literature because they were reduced to ignorance, but simply because they were forced to turn their interests to work or at least to the direct management of their possessions, activities that had previously been despised by the aristocracy.

Barbarian Invasions

Just as in the case of the alleged wave of illiteracy, there is little historical substance to the assumed devastations caused by the invasion of semi-primitive Germanic peoples. In the fifth century, the Western Roman Empire collapsed due to internal causes, primarily its economic structural limitations and its constant need to expand to acquire new slaves. The weak Mediterranean economy at this point was structurally unable to bear the burden represented by the cities and the huge Roman bureaucratic machine.

The supposed foreign invaders had in fact been integrated into the Roman world for some time. Since the third century, treaties with the Germanic border populations had been made to cope with the continual migrations, with the condition that they contributed to the military defense of the empire. Beginning with the reign of Emperor Valens (328–378), these peoples, to whom the status of *foederati* was attributed, were authorized by Rome itself to settle *intra limes*, i.e., within the borders of the empire. The collapse of the imperial structure led to the gradual increase in their autonomy until they reached full independence, with the subsequent birth of the Roman-barbarian kingdoms.

In the collective imagination, the word 'barbarian' is synonymous with 'uncivilized', but from a technological and material point of view, these were actually quite advanced civilizations. As noted by the American archaeologist Peter S. Wells, while Rome had reached an advanced state of crisis, advanced manufacturing and commercial centers flourished in northern Europe [8]. In the third century AD in Sweden, for example, the island town of Helgö was a pre-industrial center specializing in the processing of iron, gold, bronze and other metals, and produced a wide range of handicrafts. A bronze image of an Indian Buddha was found near

6.2 Four common myths about the early Middle Ages (476–1000)

the city, confirming that this settlement was connected to a vast commercial export network. Due to Roman influence, the Germanic civilizations had developed the use of money and a thriving network of exchanges with the rest of the ancient world since the end of the first century.

The contact with the Roman world gradually intensified. As early as 300 AD in Lundeborg, Denmark, there was a constant influx of high quality ceramic products and other goods of imperial origin. It was not, however, one-way traffic. The nearby town of Gudme was an advanced center for processing metals, amber, bone and horn, which were all products the Romans particularly desired. As will be seen in more detail later[3], the so-called 'barbarians' developed revolutionary technologies, such as the heavy plough, which were capable of doubling the low production volumes of the imperial age.

It is clear that these peoples had not developed a literary and legal heritage which could be compared to that of the Greco-Romans, but this is far from being the only element on which to measure the development of a civilization.

The spread of these cultures was gradual. The Romans granted the *foederati* large areas for cultivation, in which they were able to spread their technological and legal customs. The settlement of these populations was mostly planned in advance and if there had indeed been a disordered invasion of external tribes, then the settlement of the western Mediterranean would have been far more chaotic. What has traditionally – and inappropriately – been portrayed as an invasion by foreign enemies was instead a rather slow, and mostly peaceful, migratory infiltration authorized by the Roman authority itself, even if they were compelled to do so.

The Germanic population of the Visigoths who, under the leadership of Alaric, sacked the city of Rome in 410, had been among the first to be admitted within the borders of the empire. The sacking had little more than a symbolic value, because for some time the city had relinquished its functions as the political and administrative capital in favor of Ravenna and was going through a period of deep decline and abandonment. On the opposite side, the general in charge of the Roman legions was Stilicho, who was a noble Vandal just like various other prominent officers of the imperial army. In fact, rather than being a war between the Roman civilization and a foreign invasion, this was another internal conflict in an increasingly politically unstable empire, which had long since assumed a Roman-Germanic identity.

Paradoxically, the only true external invasion, that of the Huns of Attila, was stopped in 455 at the Battle of the Catalaunic Fields by a mixed army. Of course there was a Roman general, Flavius Aetius, at the head of the troops, but the Roman legionaries formed only a small minority of the rank and file. The remaining soldiers were Germanic *foederati*, whom the Romans had long since allowed

---

[3] On this subject, see section 6.6.

to settle permanently in Gaul. Among these were the Burgundians, the Alans and also the Visigoths. The final collapse came when the empire was no longer able to hold together this heterogeneous social and cultural structure. Even in 476, when the coalition of Eruli, Goti, Rugi and Turcilingi led by King Odoacer[4] ended the reign of Romulus Augustulus, it was in fact a rebellion by a general in the service of the empire for the non-payment of his troops, not the arrival of a foreign invader.

The transition from the Roman era to the early Middle Ages was a very slow and gradual one, due to the encounter between two cultures rather than a cataclysm. There was no sudden large-scale reduction of the Roman population, nor did the rich imperial literary and philosophical heritage disappear, because it was immediately recognized as superior and worthy of being handed down to posterity, precisely the point of view – already mentioned – illustrated by the phrase "*dwarves on the shoulders of giants*".

A fundamental role in this difficult encounter of civilizations was played by Christianity. The conversion of these Germanic peoples created a common ground that facilitated integration, giving rise to a new society that deeply identified itself with the common Christian ideals.

In the picture drawn by the illuminists, the only element that corresponds to actual historical facts is the major decrease in the population of the urban centers. Rome, for example, went from 500,000 inhabitants down to about 50,000 in just over two centuries. As Roger Osborne has pointed out, however, it should be remembered that in the Roman economy the cities were almost exclusively consumer centers **[9]**. Excluding the imperial bureaucratic machine, there was no trace of a modern urban economy based on services and the income of the residents often came from public donations, which in turn were funded by the taxes paid by the countryside and the imperial provinces. Cities were unproductive bubbles that were hardly sustainable even at the height of imperial economic power. As previously noted, the historian Mihail Rostovtzev effectively compared the Roman cities to the drones of a beehive **[10]**. Their decline was, in short, unavoidable, and materialized as soon as the empire manifested the first signs of weakness.

Even the loss of population in the Roman cities, because of their presumed devastation by the invaders, is little more than a legend invented by the detractors of the Middle Ages. Many buildings within the cities were torn down by the inhabitants themselves, who reused the old building material for new rural buildings, another sign of rebirth and dynamism for a world in full renewal.

In fact, it was in the Middle Ages that the first cities in a modern sense were built, which were collective and self-sufficient centers of activities and services. The best-known case is that of the Italian Communes of the thirteenth century, but there are also several cases dating back to the early Middle Ages. Examining the

---

[4] Even though he was originally a Hun or a Scirian, Odoacer had long assimilated the culture and the Roman-Germanic identity of the peoples over which he reigned.

three German border cities of Regensburg, Mainz and Cologne, Peter Wells notes that the end of imperial authority removed only the investment in new public works and in the maintenance of aqueducts and sewers [8]. At the same time, there was a proliferation of new manufacturing and commercial shops, without any reduction in the population of the inhabited area.

The collapse of trade

With the collapse of the Roman empire, the constant flow of goods and people across the Mediterranean, described by Elio Aristide in his celebratory speech (see Chapter 5) disappeared almost completely. Moreover, as is made clear from the speaker's own words, very often it was not the commercial transactions themselves that contributed to increasing the wealth of the economy as a whole, but the simple transportation of taxes and tributes to the city of Rome.

From a commercial point of view, as Enrico Artifoni observes, the Roman Mediterranean was little more than a *"sea of nobles and merchants"* and, above all, a *"sea of slaves"* [11]. The end of the central authority caused both the forced flow of tax levies and the trade in human beings to disappear. The global market in slavery, already reduced to a minimum due to the end of the empire, soon disappeared completely due to the spread of Christianity. The strong moral pressure upon the landed nobility from both the central Roman Church and the local priests led an increasing number of masters to free all their slaves to guarantee their own eternal salvation. In addition, there was an increase in the practice of mixed marriages between free men and slave women, which up to that point had been illegal, thus providing a final push towards the structural disappearance of slavery.

In the short term, this revolutionary change was undoubtedly a commercial setback and led to a drastic reduction in agricultural and craft production volumes. In the early Middle Ages, the reference productive model was mainly oriented to small-scale self-sufficiency rather than to export. At the same time, this drastic resetting of the Roman economic model provided the opportunity to begin to overcome the structural limits of the ancient world, in particular the aristocratic taboo against manual labor.

Without the contribution of the slaves, only the nobility could retain their unproductive status, and the possibility of maintaining the parasitic, idle populations of the cities vanished. Ordinary people were forced to move to the countryside: *"If they did not go out on their own to get food, they did not eat at all,"* as Anthony Bridbury summed up effectively [12].

It was a radical turning point, which Aldo Schiavone equates to a break-up of the natural course of history[5]. Over the short term, it undeniably involved some negative consequences, such as the decline of businesses and the temporary

---

[5] See Chapter 5 and Schiavone, *The End of the Past*, cit.

disappearance of the social class of intellectuals. At the same time, however, the removal of slavery laid the foundations for the technological development of the modern West, as it suddenly became necessary to search for new forms of energy that would replace, or at least complement, that of the human muscles.

The slowdown in trade transactions was a phenomenon exclusively related to the Mediterranean world. In the same period, continental Europe experienced a particularly flourishing commercial expansion, as evidenced by the aforementioned archaeological finds near the Swedish commercial city of Helgö.

The primacy of Islam

The most detrimental and misleading cliché on the early Middle Ages concerns the role of the southern shore of the Mediterranean, which had experienced the rapid expansion of Islam since the seventh century. It is a widely held opinion that while barbarians and ignorance were raging in Europe, the Islamic world was instead a thriving repository of the ancient culture, collecting and developing the legacy of Aristotle and Greek philosophy to bring it to new horizons.

While the concept of the Dark Ages is a historical inaccuracy, that of a contemporary Islamic renaissance is completely false. On the contrary, from the very beginning, the Islamic world was particularly reluctant to blend with Greek and Roman culture. Those who revived the scientific and philosophical legacy of the ancient world were almost always the *dhimmi*, or infidels, who did not practice the Islamic religion and were allowed to live in the conquered regions for the payment of a tax.

As Rodney Stark points out, citing Samuel Moffett, *"the earliest scientific book in the language of Islam was a treatise on medicine by a Syrian Christian priest in Alexandria, translated into Arabic by a Persian Jewish physician"* **[13]**.

The greatest thrust in science came mainly from the Persian world of Zoroastrian faith, which had reluctantly accepted the irrationalism of Islam. The inventor of algebra, for example, was probably a Zoroastrian priest, the Persian mathematician, astronomer and geographer Abū Ja'far Muḥammad ibn Mūsā al-Khwārizmī (about 780–850 approximately), from whose name the term 'algorithm' derives.

The great philosopher Abū'Alī al-Ḥusayn ibn'Abd Allāh ibn Sīnā, known in the West as Avicenna (980–1037), was also of Persian origin. Although of Muslim faith, he studied in the Uzbek city of Bukhara, which was already well known during the Hellenistic period. In his works, Avicenna attempted a difficult synthesis between Islam, Aristotelian philosophy and Neoplatonism, trying to reconcile the two very distant worlds to which he felt he belonged. Moreover, he dealt with medicine, mathematics and physics, also letting his love for the Greek world appear in these areas.

All the doctors of the Bukhtishu family (seventh-ninth centuries), who in the Abbasid period were significant figures in the field of medical science for over six generations, were Nestorian Christians. The translators Hunayn ibn Ishak (808–873) and Qusta ibn Luqa (820–912) were also Christians who, after the studies of al-Khwārizmī, played an essential role in the rediscovery of Greek mathematics and astronomy. The greatest astronomer in the Arab world was actually a polytheistic Sabean, Al-Battani (850–928), known in the West by his Latinized name of Albategnius.

It is therefore clear that, rather than referring to an Arabic culture, it would be correct to speak of a *dhimmi* culture. Not surprisingly, the same enlightened ruler, Al Mamun (783–833), was jokingly known as "*the commander of the infidels*", because of his openness towards philosophers, mathematicians and astronomers **[14]**.

In fact, as far as the orthodox Muslim religion is concerned, every attempt at a synthesis between faith and philosophy was bound to come up against major obstacles. Unlike Christianity, Islam never developed a theology, an attempt at the rigorous and rational description of the divine world in exclusively philosophical terms, simply because in this religion, the principle of the indescribability of God and of the inferiority of the human being is particularly strong.

One cannot pretend to know the will of Allah, nor to establish universally valid physical laws. The world is governed only by His inscrutable will, which intervenes in the course of earthly history in a much more pervasive and frequent way than the Christian God is purported to do.

If the intellectuals of the European Middle Ages had long come to a synthesis between philosophical rationalism and religious faith, following the path traced by Augustine and the other Fathers of the Church, the same cannot be said for the Islamic world, which at most limited itself to tolerating these ideas as external elements.

As Christopher Dawson observes, "*the scientific and philosophical results of Islamic culture owe little to the Arabs and Islam. It was not an original creation, but a development of the Hellenistic tradition that was incorporated into Islamic culture by the work of Aramaic and Persian men (…). Finally, in 834, under the Caliph al Mutawakkil, the orthodox reaction led to the fall of the Mutazilites who had benefited from the favor of Al Mamun and his immediate successors, and to the persecution of the philosophers. From that moment on, Islamic orthodoxy fell into a rigid traditionalism, which refused to find a meeting point with philosophy, and kept at bay all the objections of the rationalists with the formula bila kayf ('believe without asking why')*" **[15]**.

This could never have happened in the fragmented medieval Europe, even if this was the intention, because there was no unitary political power strong enough to condition the world of philosophy and scientific evolution with its directives. These are what Stark calls "*the blessings of disunity*" **[13]**.

6.3 THE LEGACY OF HELLENISTIC SCIENTIFIC TECHNOLOGY

Having cleared away some of the most common misconceptions that influence modern ideas about the Early Middle Ages, it becomes possible to see how this historical period acted as the incubation phase of the modern technological West.

In the previous chapters we have already observed how Alexandria of Egypt was the center of the first great technological revolution in Western history during the Hellenistic age, as a consequence of the passage from Greek natural philosophy to an actual form of science in the modern sense. The expulsion of the Greeks from Alexandria brought scientific innovation to a halt and began a period of deep decadence for the prestigious Egyptian school that lasted throughout the Roman period.

The last years of this long era of involution were made famous by the story of Hypatia, an Alexandrine philosopher killed in 415 AD by the Christian faction led by bishop Cyril. This episode, often exploited by those who today want to support a thesis of the violent and repressive spread of Christianity, actually had very little to do with the encounter between religions. Hypatia was a Neoplatonic philosopher and mathematician, daughter of the mathematician Theon of Alexandria. A follower of the Neoplatonic current-less polemic towards Christianity, she was very tolerant in religious matters. Many of her students were Christians, such as Synesius of Cyrene who, even after becoming bishop of Ptolemais in Cyrenaica, maintained a correspondence with her that lasted many years. Christian sources, such as the historian of the Church, Socrates Scholastic, spoke of her with great admiration. The tragic death of the Alexandrian intellectual was the result of a long conflict between Bishop Cyril (not to be confused with the more famous Saint Cyril, brother of Methodius, both patrons of the Slavic peoples, or with Cyril of Constantinople) and the Roman prefect Orestes, who was a Christian, for power over the city. The conflict led to a number of outright violent acts and disorders. What caused Hypatia's murder was not her alleged profession for feminist or secular philosophical ideals, but her political and personal closeness to the prefect. Rumors were spread which accused her of preventing a reconciliation between Orestes and Cyril, which led to her murder[6].

Cyril, who in turn would later be made a saint, enjoyed the support of Aelia Pulcheria, the sister of emperor Theodosius II, who was also Regent because the emperor was still a boy at that time. This situation gave rise to other rumors of the involvement of the imperial court in Hypatia's death. In any event, the imperial

---

[6] See Socrates of Constantinople, *Historia Ecclesiastica* VII, 15.

6.3 The legacy of Hellenistic scientific technology

court issued the decree of 416 that removed the bishop's control of the brotherhood of the parabolani who, in addition to being a voluntary health corps that tended to the care of people affected by epidemic diseases, had also become a true bodyguard of the bishop and had been accused of the murder of Hypatia. The group was placed under the direct control of the prefect. With the death of Hypatia, however, Orestes' position was weakened to the point that he left Alexandria and, from 420, Cyril remained as the only master of the city.

If we ignore the obvious symbolic value of Hypatia in the history of the emancipation of women, it should be noted that the thinking of this philosopher, often mythologized as a secular martyr of science, was actually limited to an innocuous Neoplatonic synthesis of previous philosophies. From a scientific point of view, apart from her remarkable teaching skills and from having directed some of her disciples in the construction of an astrolabe and other instruments, her contributions were marginal. The era of imperial stagnation, in short, had no exceptions.

A turning point came just at the dawn of the Middle Ages, and therefore in the years immediately following the fall of the Western Roman Empire, when John Philoponus (490–570) became the first Christian to head the school of Alexandria.

As many Fathers of the Church had done on the theological side, Philoponus tried to bring together the natural philosophy of Aristotle, Neoplatonic metaphysics and, above all, the Judeo-Christian tradition, thus unifying the great rational horizons of the ancient world under a single school of thought. These ideas had a macroscopic influence on his work in the scientific field, as for the first time there was a passage from theoretical speculation to practical experimentation.

On the basis of his direct observations, Philoponus even went as far as to disprove the undisputed authority of Aristotle, the beacon of Greek philosophy, on several points. The first of these concerned the motion of heavy bodies. By basing his reasoning exclusively on the general theory, Aristotle had claimed that heavier bodies fell faster. Philoponus, on the other hand, referred to an experiment in which bodies of different weight, dropped at the same time, reached the Earth's surface together. In short, the scholar anticipated by over one thousand years one of the most famous experiments attributed to Galileo Galilei. It is not known whether Philoponus actually performed the experiment himself – and the same is said for Galileo – but it is remarkable to find one of the statements by Aristotle contested on the basis of an experimental result.

Although Philoponus has been somewhat unfairly marginalized in the history of science, this first Christian head of the Alexandrian academy was the one who introduced scientific method into the history of science. At the very least, it is noteworthy that this revolutionary breakthrough after centuries of stagnation in the scientific world coincided with the rise of a Christian scholar at the forefront of the Alexandrian school.

Philoponus had different scientific interests. He wrote a treatise on the astrolabe, the oldest on the subject that we are aware of, and he dedicated another treatise to the propagation of light. However, his most important contributions remain those related to mechanics, revealed in his commentaries on Aristotle. According to the famous Greek philosopher, in order to maintain a body in motion it is necessary to apply a force to it, since the natural state of objects is stillness. This idea was based on the limited experimental findings he could gather, given the amount of friction present in all the machines and devices of the ancient world. However, this theory could not satisfactorily explain the motion of projectiles: how could an arrow keep moving after leaving the bow and thus after the forces acting on it have ceased to do so? Aristotle's explanation was that the air opened at the tip of the arrow and closed behind it, pushing it forward. The air would therefore not exert a resistance to motion, but would instead be a propulsive force able to push the projectile onward.

Philoponus completely overturned this concept. He reasoned that the launcher gave the projectile an impulse — which he defined as *vis cinetica* or *impetus*, and which, in some respects, is comparable to what contemporary science defines as *momentum* — and this kept the projectile moving. The question remained as to whether the impetus deteriorated over time, or whether, on the contrary, the impetus was permanent and was dissipated only by the forces of friction acting on the projectile, because sooner or later it would come to a halt. The *impetus* theory is, in some ways, a precursor of the principle of inertia, and marks a noteworthy progress with respect to Aristotelian physics.

If this interesting revival of the Alexandrian school can be attributed to the contribution of a scholar of Christian faith, then the arrival of Islamic domination in Egypt in 641 can be considered the cause of its final demise.

In the Islamic East, only the Persian philosopher Avicenna took up the theory of *impetus*, arguing further that the motion would last indefinitely in the absence of friction, and thus approaching the principle of inertia. With regard to simple machines, he also dealt with mechanisms, and used a thermometer to measure air temperature. He further claimed that light was provided by the emission of light particles, which travel at a finite speed.

However, as previously mentioned, such theories were markedly peripheral phenomena in a world that was dominated by the expression "*bila kayf*", or "*believe without asking why*". With the definitive demise of the school of Alexandria, the Muslim East experienced a scientific and cultural setback which would last for many centuries to come.

In contrast, beginning in the twelfth century, it was the scholars of medieval Europe, such as Robert Grosseteste (1175–1273), Roger Bacon (1214–1294), William of Occam (1285–1347) and Jean Buridan (1265–1361), who took over and reinvigorated the legacy of these scientific traditions.

Roger Bacon gave great importance to observation of the facts, which is why he is considered one of the founders of empiricism and, in some ways, of the scientific method. However, his work was also considered to be linked to occultism and alchemy and while it should be noted that this may appear anomalous today, at that time (and also in the following centuries) it was rather common. Some of his statements regarding a future in which self-propelled and even flying vehicles would be built are well known. These passages, and some accusations which led to him spending several years in prison, earned him a reputation as a magician and the owner of esoteric and perhaps forbidden knowledge, particularly during the Renaissance. His fall into disgrace following the death of Pope Clement IV, who greatly appreciated him, was probably due to the many enemies he had made within the Church and the Franciscan order to which he belonged, because of his intransigence in condemning the ignorance and the immorality of the clergy.

William of Occam was more concerned with theology, metaphysics and politics (he was tried as a supporter of the imperial power) than with science, but he also made an important contribution in this field with his problem-solving principle, the so-called Occam's razor, which essentially states that simpler solutions (with fewer assumptions required) are more likely to be correct than complex ones; in his words, *"Entities are not to be multiplied without necessity"*. Moreover, his work led to a reduced emphasis on the importance of Aristotelian philosophy.

Jean Buridan, who was twice rector of the Sorbonne, was of great importance in the scientific field as he settled the theory of *impetus*. According to Buridan, the *impetus* is corrupted because of resistant forces; therefore, the motion of the celestial bodies, which according to him (and contrary to what was stated by Aristotle) takes place in vacuum, continues to infinity without the need to postulate divine intelligences that keep the cosmos in motion. To refute Aristotle's theory, in which projectiles were kept in motion because of the action of air, he used the blacksmith's grinding wheel as an example, because it remained in motion while still occupying the same space. This statement was very important for two reasons. Firstly, it refuted the authority of Aristotle on the grounds of an experimental observation, and secondly because it identified translational motion with rotary motion.

Another important statement by Buridan concerned both the sphericity of Earth, which was almost taken as read given that virtually no philosopher had argued the opposite since the Greek era, but above all – and here the philosopher entered a controversial field – its diurnal rotation. Buridan is also famous for what is probably an apocryphal fable about a donkey, known as 'Buridan's ass', who dies from hunger because it cannot decide which of the two equidistant haystacks at its disposal to eat from. The apologue illustrates the concept that decision is a logical process, in which the intellect chooses between the various alternatives based on reason. In the face of two equivalent alternatives, therefore, the consequence would be a paralysis of indecision. This approach is interesting because it indicates the fundamental role given to reason by medieval philosophers.

The basis of this macroscopic migration of science from East to West was the growth of a cultural environment ever more favorable to technological innovation, an environment that allowed the development of a rationalism which was not only theoretical, but oriented towards material progress. The Christian Middle Ages was in fact the cradle of the so-called '*artes mechanicae*'.

6.4 CHRISTIANITY AND THE RISE OF THE *ARTES MECHANICAE*

Medieval humans were deeply influenced by faith in every area of their lives, and this spirituality was not limited to a simple convenient proclamation. Evidence of this can be found, for example, in the voluntary and enthusiastic support of a large number of people in the ascetic movements of the first monasticism, a choice that also involved a life of renunciation and deprivation. *"There is no doubt that this attitude has much in common with that of the great Eastern religions, but differs in a substantial way from it due to the fact that it does not lead to apathy or fatalism towards the outside world, but rather to an intensification of social activity"* **[10]**. It is in these terms that Christopher Dawson describes the pervasive influence of Christianity on the European Middle Ages, highlighting the link between religion and the deep social, material and technological evolution of these centuries.

We have already seen how Christian elaboration, in complete antithesis to that of Islam, tended to attenuate the boundaries between humankind and God. As Irenaeus of Lyon (130–202) observed, *"it is the very concept of incarnation that brings to all effects our species closer to divinity"* **[16]**. Augustine, reconsidering the myth of Eden in the light of Christian theology, observed how even Adam was immortal and perfect before the Fall. The whole of humanity would be able to return to this status as soon as earthly history reached its end, since it was the Messiah's sacrifice that cancelled the original sin. In short, humans could aspire to divinity.

For the early theologians, however, technological creations were of marginal importance. Despite the parables of Jesus and his very origins as a craftsman leading to a re-evaluation of material work, they remained as tools used only during the earthly progression of the human journey and had nothing to do with what really mattered: otherworldly salvation.

In the Middle Ages, this perspective changed radically. A first indication of this comes from the Psalter of Utrecht, an illuminated monastic manuscript made in Reims between 816 and 835. The text is accompanied by detailed ink drawings, one of which, next to Psalm 63, depicts two armies preparing for a battle (Fig. 6.1). On one side there is the army of the virtuous; on the other, the much larger army of the wicked.

6.4 Christianity and the rise of the *artes mechanicae*

Figure 6.1. The Psalter of Utrecht (816–835), Psalm 63: the wicked (on the left) and the virtuous (on the right) are preparing for battle. [Image from https://upload.wikimedia.org/wikipedia/commons/a/a2/Utrecht_Ps63_%28cropped%29_%28cropped2%29.jpg.]

The craftsmen of both factions sharpen swords. However, while the wicked have to settle on working the blade manually by rubbing it on a whetstone, the virtuous overcome them with skill by using a grinding wheel, i.e., a mechanical wheel that automates the process and yields much greater productivity. This is the first evidence of the use of this tool in the Western world and that image radically changed existing theological thinking, establishing a relationship between the refinement of technology and spiritual virtue.

This radical rethinking of material progress was expressed by Johannes Scotus Eriugena (about 815 – about 877), a theologian and philosopher of Irish origin who lived at the imperial court of Charlemagne's nephew, Charles the Bald. Eriugena coined, or at least registered in writing for the first time, the expression "*artes mechanicae*". Thanks to the aristocratic contempt for classical authors, the Latin language had no expression to describe the world of mechanics. In his eulogy of human technical achievements, Augustine himself merely referred to "*many and important arts*" **[17]**.

Even more interesting was the context in which this new expression was used for the first time. Eriugena talked about "*artes mechanicae*" in his Annotations to Marcianus, a commentary on *The Marriage of Mercury and Philology* by the Latin writer Marcianus Capella (fourth-fifth century). This great classic, widely disseminated in the Middle Ages, provided an allegorical picture of the scholastic disciplines. On the occasion of the wedding, the god Mercury offered the seven liberal arts as a gift to his wife. This formula, already designated by Seneca as the only activities "*worthy of a free man*", was free from materiality and any practical application. It consisted of the humanistic *trivium* (grammar, rhetoric and dialectic) and the scientific *quadrivium* (arithmetic, geometry, astronomy, music), the disciplines on which the entirety of academic education would be based in the Middle Ages.

In Capella's work, two arts remained excluded from the wedding gifts as unworthy: "*Apollon suggested to Jupiter that, among those he prepared, Medicine and Architecture were standing nearby, but because they care for mortal things and not the heaven and the gods of heaven, it will not be inappropriate that, if they are rejected with annoyance, they are silent in the heavenly senate*" **[18]**. These disciplines lay between theory and practical application and, in the Greco-Roman world, vacillated between the status of 'art' and the skills of vile laborers. The doctor Hippocrates was considered one of the great thinkers of antiquity and the works of the architect Fidia were universally appreciated, and yet, especially in Roman times, this separation prevailed[7].

The Christian Eriugena completely rewrote the allegory of Capella and instead included an exchange of gifts. Mercury presented the seven theoretical liberal arts, while his bride Philology offered him as many *artes mechanicae* in return, including medicine and architecture, but also metallurgy, agriculture, commerce and other traditionally despised occupations. While recognizing the difference between theoretical culture and craftsmanship, the Carolingian philosopher outlined their substantial equivalence, allowing both categories to participate in the allegorical marriage.

Eriugena also provided a theological explanation. The *artes mechanicae* were already possessed by the primitive Adam, and the recent innovations were nothing but a rediscovery of this lost knowledge. All the arts including the technological ones were, for Eriugena, "*man's links with the divine, their cultivation a means to salvation*". Technology thus ceased to be a temporary occupation concerned only with the survival of humans, and became a means to realize their ambitious return to divinity. As David Noble notes, Eriugena "*believed that through practical effort and study, mankind's prelapsarian powers could be at least partially recovered and could contribute, in the process, to the restoration of perfection*" **[19]**.

The thinking of the Carolingian theologian deeply influenced medieval civilization. The rediscovery of the dignity of the *artes mechanicae*, initiated by Eriugena, continued with the bishop Hugues de Saint Victor (about 1096–1141) who, reversing every previous prejudice, dedicated several pages to a detailed description and classification of *artes mechanicae*.

---

[7] The only exception to this was the historian Marcus Terentius Varro (116 BC – 27 BC), who included both architecture and medicine among the *artes liberals*. The majority of authors prefer to exclude them as well.

A final push came from the apocalyptic movement of millenarianism, which revived the expectation of the second coming of Christ. Among the leading exponents of this movement was the ascetic Joachim of Fiore (about 1130–1202), who was consultant to three popes.

At first, the link between millenarianism and technology seems unbelievable, as it has previously been noted how the expectation of an imminent return of Christ on Earth inhibited the re-evaluation of craftsmanship and matter in the first Christian centuries. Only the postponement of this prophecy to a less immediate future led the first Christians to deal with earthly matters[8].

After the theological turning point of Eriugena and Hugues de Saint Victor, millenarianism regarded exercising and innovating the *artes mechanicae* in an extremely positive way, precisely by virtue of their spiritual nature.

It was not by chance that the point of convergence between the apocalyptic followers of Joachim of Fiore and this new philosophy was represented by a man of science, Roger Bacon. According to this Franciscan scholar, technological innovation could even be seen as a tool to accelerate the imminent coming of the final times and to bring humans back to divine perfection.

As Noble summarizes: "*If Bacon, following Eriugena and Hugues of Saint Victor, perceived the advance of the arts as a means of restoring humanity's lost divinity, he now saw it at the same time, following Joachim of Fiore, as a means of anticipating and preparing for the kingdom to come, and as a sure sign in and of itself that that kingdom was at hand*" **[20]**. With this last conceptual turn, Bacon actually freed humanity from any kind of fatalistic standpoint, making them definitely masters of their own destiny. The faster the technological progress could be achieved, the more the Millennium and the final perfection would accelerate. Technology would now become one of the keys to achieving Christian virtue.

6.5 WORK AND PROGRESS, THE TURNING POINT OF MONASTICISM

The medieval chronicles reveal that Eriugena chose to spend his old age in a Benedictine monastery. The anonymous illustrator who had drawn the clash between the impious and the virtuous on the Psalter of Utrecht also belonged to the same order **[21]**. Gioacchino (Joachim) da Fiore was instead a Cistercian abbot while, as mentioned earlier, Roger Bacon was a Franciscan. This list could go on for some time, as almost all the initiators of the medieval technological revolution belonged to the world of monasticism.

---

[8] See the previous section 4.6 and the reference to 2 Thessalonians.

Yet this social phenomenon was born in Egypt in the third century from the most radical current of the Christian religion which, even under the influence of Gnosticism, despised matter, society and progress and believed that the only possible solution to reach holiness was to "escape from the world".

Jesus himself had spent forty days in the desert of Galilee fighting the temptations of the flesh, and prior to that, John the Baptist had left his social life to live a life of pure spirituality. In the Gospels, he is described as *"dressed in camel hair, with a leather belt around his hips"* [22].

The first monks were inspired by these examples, in particular Antony, who lived in Egypt between 251 and 356 AD. His ideal of life was asceticism, or spiritual improvement which is achieved by completely abandoning all links with earthly needs. This was a strictly individual way of life. The Greek word *"monakos"*, moreover, actually means *"one who lives in solitude"*.

How was it possible that from such radical assertions, European medieval monasticism could become the leader of technological innovation and material progress?

The architect of this epochal turn was Pachomius (292–348), an Egyptian monk whom William Harmless defines as a true *"genius of organization of the ancient world"* [23]. Initially, Pachomius had chosen to live in the desert in solitude following the example set by Antony. The chronicles report, however, that in the desert he soon felt the urge to turn his teaching to others, creating a lifestyle at the same time strongly isolated from rest of the world and based on the construction of a new cohesive and collective society.

Under his guidance, in the surroundings of Thebes, the *Koinonia* was started, a hierarchically organized federation subdivided into eleven monasteries that gathered about 5,000 men and women. As Harmless observes, *"until the Middle Ages, with the rise of Cluny and the Cistercians, no monastic order could equal that of Pachomius in terms of size and organizational complexity"* [24].

Their network of villages was dedicated to meditation and spiritual refinement but, wanting to keep their distance from the rest of the world, they were also forced to dedicate themselves to economic self-sufficiency. Each monastery was surrounded by walls to protect itself from external threats. Inside, there were barns and warehouses but also, most notably, many craft shops, ranging from simple basket makers to carpenters, tailors, shoemakers, bakers and blacksmiths. The *Koinonia* was a veritable industrial society, *ante litteram*, with a careful subdivision of production processes and a considerable degree of specialization to its components. There were even some guilds, with common residences for those who worked in the same sector. The historian Palladius reported with amazement that *"in the monastery of Panopoli, strong of three hundred men, I counted fifteen tailors, seven blacksmiths, four carpenters, twelve camel-drivers, fifteen weavers"* [25].

6.5 Work and progress, the turning point of monasticism

What is even more surprising is that, for Pachomius and his followers, these fervent productive activities were not merely considered a necessity, as they had been for Antony, but were part of the monastic life and of the path toward the spiritual improvement of the individual. In the Rule of Pachiomus, manual labor had an important place alongside the need to learn to read and write, and that of meditation and assembly prayer. Pachomius wrote that *"the brothers are not forced to work excessively but a moderate activity spurs everyone to work"* **[26]**.

Manual work was one of the mandatory steps towards achieving the spiritual improvement of the monk and it was how, in the sands of the Egyptian desert, a cultural turning point occurred which would then be described by Eriugena in Europe more than five hundred years later.

Once again, there was a pattern of cultural migration from East to West. Even the spread of Egyptian monasticism, as with the Alexandrian school, suffered a final setback with the arrival of the Arabs and Islam. In the meantime, however, Jerome had translated the Rule of Pachomius (404) into Latin, also introducing to the West this radical alternative to the pagan ideal of the *otium*, which actually ended up becoming a synonym for vice and corruption with the passing of time. The one who took up the legacy of Pachomius was Benedict of Nursia (480–547), *"the creator of Western monasticism par excellence"* **[27]**.

Referring to the Pachomian model, the Benedictine rule introduced the ideal of *ora et labora* (pray and work) as the perfect synthesis of the life of the monk. As had happened in Egypt, the revaluation of work as an instrument of spiritual perfection in the West coincided with the development of a strongly structured economic and organizational model.

As observed by Anna Rapetti, in Europe *"the Rule represented the triumph of the organization and of the institution over the spontaneity that had characterized to that point the cenobitic life, based mainly on personal ties and on the charisma of the founder"* **[28]**. The spread of Benedictine monasticism received a decisive boost with the rise of the Carolingian empire, which became its official sponsor. Charlemagne and his son Ludwig even imposed the Benedictine order on all the religious communities subjected to their authority. During this time, the monks were free to spread this new attitude towards work across Europe and its strongest symbol is the illuminated gospel of Winchester, in which God is depicted as an engineer holding in his hands a scale, a pair of compasses and a carpenter's square.

There was one caveat. For both for Pachomius and for Benedict, the total dedication to work as an end in itself was also incorrect behavior, just like *otium*. A life exclusively devoted to practical activities was in fact an empty existence, as it was one devoid of prayer, contemplation and spirituality. According to Ernst Benz, *"the theorists of monasticism favored manual work only as a means to a spiritual end"* **[29]**. Physical fatigue was therefore strongly discouraged, as it absorbed the energies that had to be devoted above all to the care of the soul.

This was one of the main reasons why the monks were always at the forefront of mechanical and technological innovation. Each tool allowed human beings to reduce their physical efforts, thus maintaining the balance between work and prayer. The Benedictines, for example, were pioneers of the use of wind and water mills, heavy ploughs and new agricultural techniques. In the field of music, they were the first to create and build a pipe organ, in Winchester, one of the most complex mechanical instruments of medieval Europe.

6.6 THE MEDIEVAL TECHNOLOGICAL REVOLUTION

In spite of the myth of the Dark Ages, it was in the medieval era that the second great technological revolution in Western history began. As is evident, it was a silent turning point, bound to have a lesser literary success than that of the previous Hellenistic revolution. Yet its actual impact on daily life was infinitely larger, to the point that it can reasonably be said that modern civilization was born in these centuries, and with it, the current technological and scientific primacy of the West.

With respect to the Hellenistic age, the change in perspective is radical. While the technological innovations of the Alexandrians were essentially *mirabilia*, that is, entertainment products intended primarily to arouse amazement in intellectuals and to decorate noble residences, the medieval ones were very concrete applications, which were not only available at the lower social levels but were of immediate economic use. Their goals, moreover, were to replace the muscle power of the slaves, save time and optimize the resources dedicated to work, particularly in the monastic environment.

It is not surprising then to read the words of an anonymous peasant living in around the year 1000, which describe the first major technological breakthrough that occurred in the opening centuries of the Early Middle Ages, a short time after the dissolution of the Roman Empire: *"I work very hard. I go out from home at dawn, I lead the oxen into the fields and harness them to the plough. There is no winter so hard to keep me at home idle"* [30].

The technological device with animal traction, described by the farmer, bore little resemblance to the plough previously used in Roman times to till the fields. According to the American historian Lynn White Jr., this invention had such a deep impact on daily life that it also triggered the philosophical turn towards work discussed in the previous chapters [31]. The impact on the collective imagination was actually macroscopic. As its use became more widespread, the iconography of the monastic calendars began to idealize the figure of man as master and transformer of nature.

The previous Greco-Roman plough was a very simple device from which only a wooden spike protruded, which traced a narrow slit as it passed over the ground in which the seeds could be laid for cultivation. The medieval heavy plough, or

6.6 The medieval technological revolution

asymmetrical plough, was a much more elaborate mechanical device equipped with wheels and bearing a shaped iron blade that made a much deeper slit in the ground, as well as a large metal plough that raised the ground with the slit, cutting it horizontally. There was also an oblique metal ear, which turned the clods after the passage of the plough [8]. Unlike the Hellenistic technology, which remained almost completely devoid of economic application, this first technological innovation had an immediate effect on the whole of medieval society. The instrument, in addition to significantly increasing agricultural yield, greatly reduced the time required for ploughing. For communities that aimed at self-sufficiency, it freed many people from the primary sector, thus laying the foundations for further technological developments.

However, what is most surprising is that the invention of the heavy plough was introduced by those same Germanic peoples that the illuminist historiography imaginatively depicted as nomadic and semi-primitive hordes. In short, far from being a setback, the dissolution of the Roman Empire was a new beginning for Western technology after centuries of stagnation.

The rigid collar, which made it possible to replace oxen with horses for the heavy plough and thus made agricultural operations even quicker, was also of 'barbarian' origin. Combined together, the two innovations produced a leap forward that was hitherto unimaginable. In addition, the three-year crop rotation was introduced somewhere around the year 1000 and while this was not a new technology in the strictest sense, the alternation of cereals, legumes and fallow was a rational and innovative approach, a sign of strong organizational dynamism in agriculture.

As previously mentioned, windmills and water mills spread throughout Europe thanks to Benedictine monasticism. In the census commissioned in 1086 by King William the Conqueror[9], more than 6,500 water mills were listed in England, one for every 50 families.

Architecturally, the large Romanesque churches and especially the Gothic cathedrals of the thirteenth century are sufficient to illustrate the new technological skills of medieval builders, creators of monumental works of extreme complexity which defied gravity and demonstrated an extraordinary mastery of structural equilibrium.

As for crafts, in addition to the aforementioned introduction of the rotating grinding wheel for automatically sharpening swords, the most complete and exhaustive historical testimony on the many novelties of the period dates back to the twelfth century and comes once again from the monastic environment. The treatise *On the various arts*, compiled by a Benedictine monk named Teofilus, describes in great detail the proliferation of new techniques for working materials,

---

[9] This record is known as "The Domesday Book".

illustrating various technologically sophisticated operations in the smallest details, such as working glass, gold and gems, glazing, painting on walls, on wooden boards and on parchment, metal smelting and tinning. Unlike what happened in the Greco-Roman world, where the few authors on technology limited themselves to writing inaccurate and summarized descriptions because they had not mastered the technologies they were concerned with, Teofilus demonstrated his direct experience in these disciplines and wrote a true operating manual.

6.7 FINANCE AND ACCOUNTING IN THE MIDDLE AGES: THE DAWN OF THE MODERN ECONOMY

In examining the fate of the Hellenistic technological revolution, it has already been observed that these great scientific achievements could not be translated into practical technology because of an archaic economic system which was totally unprepared for an industrial evolution. In this regard, the Christian Middle Ages mark an epochal turning point in the history of the West, because in addition to laying the foundations for overcoming the millennia-old prejudice against practical work, it produced two innovations essential for the birth of modern economy, leading to the rise of capitalism and the systematic application of technology in production processes. These innovations were the birth of the financial market and the creation of an organic and rational system of economic description and planning: double-entry accounting.

With regard to the world of finance, it has been noted that banking institutions existed as far back as the Greek times. However, these institutions almost never extended credit to craftsmen and merchants, preferring to avoid being 'contaminated' by humble and undignified sectors, even if this meant losing profitable business.

With the advent of the great monotheistic religions this practice was disregarded, but there was also strong hostility towards the concept of charging interest on loans, an essential element for the growth of a money market. The Old Testament in fact, prescribes that *"if one of your brothers falls into poverty (...), you will not lend him money to interest, nor will you give him food to usury"* **[32]**. In another passage, it is explicitly clarified that *"brothers"* means those belonging to the Jewish people: *"if you lend money to any of my people, to the poor who is with you, you will not treat him as an usurer; you will not ask any interest from him"* **[33]**.

For the Jews, there was therefore no limitation to the practice of banking with the so-called *"gentiles"*, or people of different religions. In contrast, both Christianity and Islam extended the prohibition to the whole of humankind, effectively condemning the request for interest as sinful. The Middle Eastern mentality equated any request for *superabundantia*, or a surplus with respect to the capital lent, to a form of extortion.

This was a typical approach of pre-capitalist economies, which did not attribute an economic value to the factor of time. From this point of view, loaning money did not cause a reduction of the capital, nor did it require any physical or intellectual work worthy of remuneration by the lender. From a modern perspective, during the period the money has been loaned out the lender renounces the possibility of investing these resources into a productive activity and thus making a profit. This is why the lender must be rewarded with a payment of interest.

In antiquity, however, this type of reasoning was absolutely inconceivable. Both in Judeo-Christian Palestine and in Greece or Rome, most of the monetary capital accumulated by the nobility was unproductive and was kept back purely for their own well-being. The idea of reinvesting a substantial part of these sums into a new economic activity was not considered, so any request for a payment of interest was automatically equivalent to usury.

The Christian ban on requesting a remuneration for loans derived from the evangelical command of *"doing good and lending without any hope"* [34]. Undeniably, it had a very strong impact on the economic development of the early Middle Ages. The Diaspora Jews filled the structural vacuum and soon specialized in lending money at high interest.

At the same time, however, the experience of European monasticism introduced a first form of capitalist logic in the Christian context. Having made a vow of separation from the earthly world, the Benedictines and the Cistercians could not enjoy the wealth accumulated by their orders. As a result, they systematically reinvested it into their agricultural and craft activities, creating ever more flourishing economic centers.

In addition to revolutionizing the world of production and technology, this type of organization also influenced the concept of money. For medieval religious orders to lend money meant impoverishing the growth cycle of the monastery, and it soon became clear that the evangelical notion of a free loan needed to be systematically rethought. This need became even more evident when there was a strong surge in demand. It is known that after the year 1000, there was a rapid spread of commercial activities, and the bourgeois who embarked on this type of project almost always needed money to keep them going.

While Gratian (about 1075 – about 1145), the father of canonic law, continued to sanction the prohibition of every form of loan at interest in his *Decretum*, a first important breakthrough came from Thomas Aquinas (1225–1274), who drew a line between so-called *"lawful interest"* and usury. According to the Italian theologian, any loan in fact implied for the lender:

- the emergence of damage (*"damnus emergens"*), deriving from the dispossession of one's own goods;
- the cessation of a profit (*"lucrum cessans"*), recognizing the impossibility of reinvesting the sum loaned;
- the risk of insolvency of the loanee (*"periculum sortis"*).

In this reasoning, the existence of a surplus, with respect to the capital made available, was therefore lawful and justified, at least to the extent that it covered these three damages.

This was therefore a first concession, dictated by the need for Thomas to mediate between the official ecclesiastical ruling and everyday reality. According to the Church's thinking, deeply influenced by the theories of Aristotle, money was not an independent commodity, but only a unit of measure of the value. From this point of view, its growth due to the effect of circulation alone remained an unnatural and essentially unfair act. In everyday life, however, the need to compensate for the temporary unavailability of the capital dominated, and the practice of interest-bearing loans became widespread.

In *canto* XI of *Inferno*, Dante Alighieri placed those who lent money at interest among the *"violent against nature"*, together with the sodomites and the inhabitants of the Occitan city of Cahors, famous for the practice of interest loans. Despite the condemnation of Dante, still based on the Aristotelian conception of the sterility of money, this type of activity proliferated throughout Europe. The Lombards, a generic term used to define the bankers of northern Italy in the Middle Ages, were even more famous than the inhabitants of Cahors.

The Roman Curia itself, under the pontificate of Gregory IX (1227–1241) and Innocent IV (1243–1254), had officially recognized the Lombards and awarded them the task of collecting ecclesiastical taxes. In France and in the rest of Europe, the Lombards even became known by the nickname of "usurers of the Pope". The appellation, clearly derogatory, derived from their practice of applying a very high rate on mortgages, which ranged from 40 to 70 percent (in spite of the 'lawful interest' outlined by Thomas Aquinas).

The first direct involvement by ecclesiastical orders in the banking practice was due to the order of the Templars. Warrior monks, but also brilliant administrators, the Knights of the Temple had begun their own financial activity, depositing and keeping custody of the assets of those who left on pilgrimage for the Holy Land. Having several houses (*magioni*) throughout Europe and the Crusader Kingdoms at their disposal, and having accumulated a great wealth because of the integral vow of poverty required of those who aspired to join the order, the Templars began to carry out this type of activity with the blessing of the Pope. Any pilgrims, merchants, or even commanders who wanted to transfer their wealth far away without running the risk of taking the money with them during the trip simply deposited these sums at a Temple site and withdrew them with a simple letter of credit at whichever Templar house was nearest to their destination.

For their part, the Templars rigorously documented all the financial transactions with state-of-the-art accounting, thus preventing anyone from seizing the letter and retrieving the money at their destination. One fragment of these records, consisting of only eight sheets of parchment relating to the years 1295 and 1296, shows 222 accounting records in a fixed pattern, containing the amount, the name of the depositor, the cause, the name of the future recipient of the credit, the date, the name of the cashier and a note of the connection between the register and the receipt [35].

With regard to the actual loans, the order managed to circumvent the prohibition of charging interest by collecting the surplus with the application of a particularly favorable exchange rate between one currency and another. It was not in any way fraudulent. As mentioned, the Christian Middle Ages had largely overcome the anachronistic prohibition against the remuneration of capital in daily practice, and everybody now accepted that any loan should be repaid with an extra amount, albeit in the most disparate forms.

This was particularly evident from the difference between the Christian elaboration, which produced interpretations of the sacred text capable of departing, even radically, from its literal meaning; and the Islamic one, which for a long time remained anchored to the integral meaning of the provisions of the Koran.

As Richard Barber observes, in the Templar reports of the financial operations between the years 1290–1293 involving the French crown, "*the payment of interest on loans is evident, although it is not always openly stated in the documents that record the transactions*" [36]. Alternatively, in the case of a mortgage loan, the Templars used to collect the produce derived from the cultivation of the lands which was given to them as collateral [37].

The Christian openness to the financial market became particularly evident in the fifteenth century, with the establishment of the first *Monti di Pietà* (pawnshops)[10] by the friars of San Francesco, or *Osservanti Minori*. These ecclesiastical institutes were born with the explicit intention of financing the "*pauperes pinguiores*", the "richest among the poor", which consisted of those craftsmen and shopkeepers that had been systematically ignored by credit institutions since the time of ancient Greece.

Thus, for the first time in history, the financial market could support the economy, allowing companies to anticipate what resources would be necessary and accelerating the growth of the system. Initially, some *Monti di Pietà* were there to provide assistance and did not charge any interest. They coexisted alongside others which instead applied 'lawful interest' under the terms identified by Thomas Aquinas.

The latter remained highly controversial, until the issue was definitively resolved by the Papal Bull (public decree or charter) *Inter multiplices*, issued on May 4, 1515 by Pope Leo X at the conclusion of the work of the tenth session of the fifth Lateran Council. The pontiff, while recognizing the moral primacy of non-profit subjects, also declared that the work of the ecclesiastical *Monti di Pietà* that charged interest to be lawful and meritorious, since this income was necessary

---

[10] The first *Monte di Pietà* was founded in Perugia on April 13, 1462. "Monte", as in mountain, in this case means "capital".

to provide for internal management expenses. Moreover, the Bull encouraged the opening of new institutions of this type, at the indulgence of the Holy See. After this provision, which put an end to the debate regarding charging interest under penalty of excommunication, the presence of institutions which tried to increase their treasury by obtaining a profit, in addition to covering internal management costs, was also accepted. This was the dawn of modern finance and of the future advent of capitalism.

The Medieval period also saw the birth of a new complex accounting system for commercial activities, the so-called 'double entry', consisting of the dual financial and economic registration of company dealings. Prior to this, only financial entries, i.e. cash receipts and payments, had been kept in the account books, whereas this new system also recorded the economic aspects (revenues and costs) in order to determine whether the company was being managed at a profit or a loss, regardless of the immediate monetary trend of receipts and payments.

During the Renaissance period, Luca Pacioli (1447–1517), a Franciscan mathematician and humanist, and a friend of Leonardo da Vinci, described this system in his *Summa of arithmetic, geometry, proportion and proportionality*. However, it is now accepted by all scholars that his was just a popularization of a practice already established in the Middle Ages.

It is difficult to attribute the precise location where this technique began, but it was probably of Italian origin, as documented analytically by Carlo Antinori [**38**]. In Venice and in Milan, in Genoa and in Emilia Romagna, autonomous accounting calculations had flourished since the fourteenth century and these led, after the various systems had been compared, to the systematic development of this convention.

The first surviving document drawn up in double entry was compiled in Roman numerals by the *Compagnia di Genova* and is dated to 1340. This was the golden age of the Communes, of the Maritime Republics, and of the large commercial and artisan guilds of the city, the natural outcome of the early medieval evolution.

Compared to the great philosophical, technological and financial revolutions described so far, the introduction of this accounting tool may seem little more than a detail for specialists, but this is not the case. As Luca Pacioli observed, this new convention, while maintaining the original function of "*taking note*" of the business events that had already occurred, above all had the objective of providing a starting point for planning a "*Wise and honest administration in the future*" [**39**].

The task of those who managed a company, therefore, was no longer merely to record what had happened, but also and above all to plan for the future, based on past and present data. According to Pacioli, "*the reason for the present lies, in fact, in the past,*" and "*the past can help us to look into the future*" [**40**].

With double-entry accounting, a rationalistic and planned structure to the economy developed, in which the successful no longer limited themselves to passively enjoying the income of their own companies, but pre-planned their progress, starting with the analysis of the available information. With this new strategic approach, which for the first time considered the constant monitoring of business as a priority, the logic of capitalist reinvestment of profits, and of investment in technological instruments, made its way into administration. Machinery now ceased being used as a pure form of entertainment and was instead designed and planned in advance as assets at the service of the economic cycle.

References

1. Voltaire, *Complete works*, vol. XII, quoted in Stark R., *Bearing False Witness: Debunking Centuries of Anti-Catholic History*, Templeton Press, West Conshohocken 2016.
2. *Ibid*, p. 108, n. 6.
3. Gibbon E., *The History of the Decline and Fall of the Roman Empire, vol. 6*, Strahan & Cadell, London 1776–1789.
4. Russell B., Foulkes P. (ed.), *Wisdom of the West,* Macdonald, London 1959.
5. Manchester W., *A world lit only by fire: the medieval mind and the Renaissance, portrait of an age*, Little Brown & C., New York 1993.
6. Stark R., *Bearing False Witness: Debunking Centuries of Anti-Catholic History*, Templeton Press, West Conshohocken 2016.
7. Harris W.V., *Ancient Literacy*, Harvard University Press, Cambridge, Massachusetts, 1989.
8. Wells P.S., *Barbarians to Angels: The Dark Ages Reconsidered*, W.W. Norton, New York 2008.
9. Osborne R., *Civilization: a new history of the Western world*, Pegasus, New York 2006, p. 163.
10. Dawson C., *The Making of Europe: An Introduction to the History of European Unity*, Sheed and Ward, London 1932, reissued by the Catholic University of America Press, Washington DC 2003.
11. Artifoni E., *Storia medievale*, Donzelli, Rome, Italy 2003, p. 61.
12. Bridbury A.R., *The dark ages*, in "The Economic History Review", n. 22, 1969, p. 533.
13. Stark R., *How the West Won. The Neglected Story of the Triumph of Modernity*, Intercollegiate Studies Institute, Wilmington, Delaware 2014.
14. Reference 10, p. 235.
15. *Ibid*, pp. 236 and 241.
16. Quoted in Weber M., *Economy and society*, University of California Press, Oakland 1978.
17. Augustine, *The City of God* XXII, 24.
18. Martianus Capella, *On the Marriage of Philology and Mercury*, IX, 981.
19. Noble D. F., *The religion of technology: the divinity of man and the spirit of invention*, New York and London 1997, p. 17.
20. *Ibid*, p. 26.
21. Reference 19, p.13
22. Gospel of Mark 1, 6; cfr. Matthew 3, 4.
23. Harmless W., *Desert Christians. An Introduction to the Literature of Early Monasticism*, Oxford University Press, Oxford 2004, p. 115.
24. *Ibid*, p. 13
25. Palladius, *Lausiac History* 32, 9.
26. Pachomius, *Preacepta ac leges* 3.
27. Rapetti A., *Storia del monachesimo occidentale*, Il Mulino, Bologna, Italy 2013, p. 38.

28. *Ibid*, p. 39
29. Quoted in Reference 19, p. 13.
30. Benham A., *English Literature from Widsith to the Death of Chaucer*, Yale University Press, New Haven 1916, p. 26.
31. White Jr. L., *Cultural Climates and Technological Advance in the Middle Ages* in Viator 2 (1971), pp.171–202.
32. Book of Leviticus 25, 35.
33. Book of Exodus 22, 25.
34. Gospel of Luke 6, 35.
35. Barber R., *The Trials of the Templars*, Cambridge University Press, Cambridge 1993.
36. *Ibid*
37. Chianazzi P., *Gli ordini cavallereschi: storie di confraternite militari*, Edizioni Universitarie Romane, Roma, Italy 2013, p. 173.
38. Antinori C., *La contabilità pratica prima di Luca Pacioli: origine della partita doppia* in De Computis, 1 (2004), pp. 4-23.
39. Ciambotti M., *Finalità e funzioni della contabilità in partita doppia nell'opera di Luca Pacioli*, in Cesaroni F.M., Ciambotti M., Gamba E., Montebelli V., *Le tre facce del poliedrico Luca Pacioli*, Quaderni del Centro Internazionale di Studi Urbino e la Prospettiva, Arti Grafiche Editoriali, Urbino, Italy 2010, pp. 11-25.
40. *Ibid*, p. 18.

7

The beginning of scientific technology

7.1 TECHNOLOGY AT THE TRANSITION FROM THE MIDDLE AGES TO THE MODERN AGE

The first attempt to implement a 'scientific technology' in the Hellenistic age was aborted with the return of traditional empirical technology, before a revolutionary turnaround in the Middle Ages which introduced many innovations into everyday life.

The end of the Middle Ages is conventionally deemed to coincide with the discovery of America, towards the end of the fifteenth century. The period saw a wealth of knowledge and innovations spread throughout Europe and it established a clear primacy of the West over the rest of the world. Some of the major achievements that characterized the beginning of the modern age are listed below, although the evolution of many of these also dates back to the Middle Ages:

- Movable metal type printing. Sales of the first printed book, the Bible, began in 1455. Johannes Gutenberg (1398–1468) founded a company and developed a new technology by combining existing technologies. He had to develop the technologies to cast the movable metal type, study the composition of a suitable oil-based ink rather than the water-based ones in use at that time, and build a printing press. The introduction of his press, which was inspired by the presses used in the winemaking industry of the Rhine region, brought in the possibility of making books available in quantities previously unthinkable, at relatively low costs.
- Spectacles. Apart from questionable, or at least exceptional precedents, in particular the emerald that Nero used to watch circus shows, spectacles were introduced in the fourteenth century, having probably been invented

by a friar from a Venetian convent. At the beginning they were rare and expensive, but towards the end of the fifteenth century they became sufficiently widespread to begin to have a significant economic impact. In fact, they extended the working life of all those whose jobs required good visual ability, who would otherwise have had to stop working at around 40–50 years old, the age at which the presbyopia (long-sightedness) tended to manifest.

- Low cost glass windows. Glass windows came into use in the Roman period, for religious buildings, some public buildings such as the baths, and occasionally in the houses of particularly well-to-do people. In normal houses, without glass, the occupants used to use canvas or translucent paper to allow in light while protecting themselves from the cold. At the end of the Middle Ages, windows equipped with glass began to appear even in normal buildings, although they were still limited to the houses of people of high social standing. By about this time, the Roman hypocaust and the central hearth had been almost completely replaced by the wall fireplace, which had spread into virtually all homes, clearly showing that the general standards of comfort of the population had increased.
- Mechanical watches. The first mechanical clocks, which were tower clocks generally installed on church towers, were built in 1200. Their precision was limited, and they were generally regulated every day at 12 o'clock. By the end of the fifteenth century they had become widespread and the first portable mechanical watches began to be produced, to be worn hanging from a chain like pendants. These were clocks built with only the hour hand and they therefore had limited precision, to about a quarter of an hour.
- Use of draft animals. As previously stated, horses began to be properly harnessed towards the end of the tenth century, allowing the animals to exert a much stronger force than before. More or less at the same time, the practice of using permanent horseshoes became widespread. This led to considerable advantages both in agriculture and in land transportation. Over the following centuries these practices spread, alongside considerable improvements in the vehicles as carriages, commonly equipped with steering mechanisms and sometimes suspensions, began to be used.
- The compass. Like gunpowder and printing, this was likely a Chinese invention. As Francis Bacon commented explicitly, although he was probably unaware of their common geographical origin: *"The press, gunpowder and the compass: these three inventions have completely changed the face of the world and the state of affairs everywhere; the first in the literature, the second in war, the third in navigation; they have brought so many changes, so deep, that no empire, no sect, no star seems to have exerted as much power and influence in the life of man as these mechanical discoveries"* **[1]**.

Karl Marx also claimed that these three inventions were the basis of bourgeois society. However, in all three cases, as well as paper, the fourth invention attributed to the Chinese, their effects were felt much more in the West than in China.

Together with these typically medieval inventions, which became widespread at the end of the Middle Ages, there are two innovations that revolutionized military technology:

- Portable crossbows. As already mentioned in examining Roman military technology, the Greek *ballista* was perfected and widely adopted by the Romans. The crossbow, the small-scale version of this weapon that could be transported by a single soldier, was developed later, with its maximum use between the eleventh and the fourteenth century. Able to pierce a metal armor from a considerable distance (second in power only to the composite arc, a rare device that required great skill from its operator), the crossbow led to an increase in the deaths on the battlefield and was banned by the Lateran II Council in 1139, with a Papal Bull later renewed by Pope Innocent II. The prohibition held for battles between Christian armies, although since it could not be applied to Muslim armies and heretics, it was permitted against non-Christians. Regardless, the crossbow radically changed the strategy and the way of fighting and the ban was almost never applied in practice.
- Gunpowder and firearms. Gunpowder was probably invented in China, but it was certainly used in Europe in the eleventh century. In order to use it as a propelling explosive, i.e. for the construction of firearms, it was necessary to make considerable progress in metallurgy and in metal forging, which were required for making barrels able to withstand the great pressures that are produced when firing. The late fifteenth century saw the availability of portable firearms and artillery that was capable of working safely. In the beginning at least, firearms were met with several ethical reservations, but they were accepted more quickly than crossbows, perhaps thanks to that precedent. Firearms, particularly heavy ones, led to actual revolutions both in military strategy and in the construction of fortresses and ships. Firearms were one of the first technologies in which a very small action requiring a very limited force could have great consequences. In addition, a firearm could be operated involuntarily and therefore required measures to prevent such an occurrence.

It should be noted that the innovations mentioned above, and the many others that accompanied them, did not require true scientific technology, as had happened with the innovations introduced during the Hellenistic period. What remained of the texts on geometry, optics, hydraulics and pneumatics that had been handed

128 The beginning of scientific technology

down for centuries was more than enough to implement these devices. They were perfected more than anything else by proceeding by trial and error, in the tradition of pre-scientific technology.

7.2 THE "THEATRES OF MACHINES" OF THE RENAISSANCE

In the classical world, the professional figure entrusted with the planning and execution of buildings, as well as the design and implementation of the machines used at the construction site, was the architect (*architectus*).

Towards the end of the Roman period, the *architectus* was joined by another figure, that of the *ingegnarius*, whose task was the conception and realization of the *ingenia*, or machines, mostly siege engines or other military devices. In the Middle Ages and then again in the Renaissance, the two roles were often combined in one person, who was both an architect and a military engineer, with the distinction only becoming important between the end of the seventeenth and the beginning of the eighteenth century.

The great architects of the fifteenth century, like Leon Battista Alberti (1404–1472) or Filippo Brunelleschi (1377–1446), were 'complete artists', able to turn their hand to multiple fields of knowledge, ranging from art to science and technology.

Many of them left works that were both didactic texts and sketchbooks for their personal use, and were sometimes valuable books intended for a wider audience. These texts, which became known as "theatres of machines", became a kind of literary genre that received a wide circulation with the invention of the press.

The initiators of this trend were architect Villard de Honnecourt (1200–1250) and medical doctor Guido da Vigevano (1280–1349) who were both interested in machines, from those used on construction sites to military machines or automata. Their books contain drawings which are often quite rough and difficult to interpret.

The sketches within the "theatres of machines" attempted to represent the three-dimensional reality of the mechanism under examination, but with no attempt to provide precise information on the dimensions of the various parts and no ambition to be real construction drawings. The construction drawing is a plan used to link the designer who both conceives the machine and defines all its dimensions, with the builder (the worker, as will be defined during the industrial revolution) who has to realize and then assemble the various parts without taking any decision about its design. At this point, that link was still to be forged.

For example, in providing the description of a mechanism such as his wind wagon, the first self-propelled vehicle described in history, Guido da Vigevano provided some approximate dimensions but then noted that an expert windmill builder would be required to define all the details. Today, for a scientific

7.2 The "theatres of machines" of the Renaissance

technology, the designer of such machinery would establish, for example, the transmission ratios of the various pairs of gear wheels, the dimensions of the windmill sails, the radius of the wheels, and the other fundamental dimensions. Back then, the craftsman had to determine them according to his experience. A notebook page by Villard de Honnecourt, showing machines including a water-powered saw and a dove-shaped automaton, is shown in Figure 7.1.

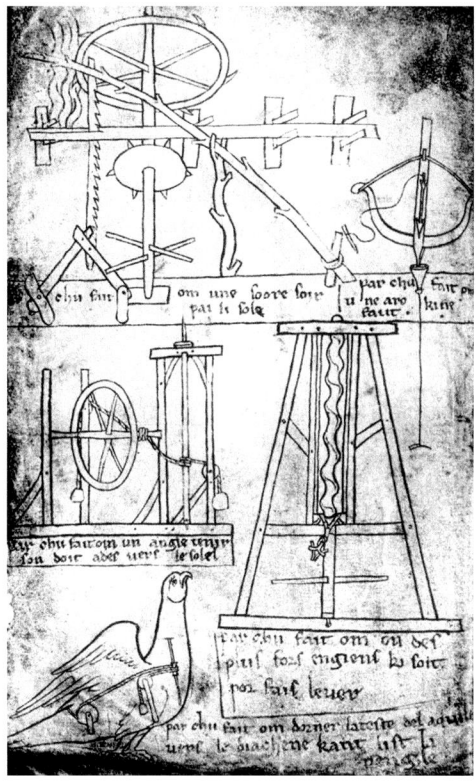

Figure 7.1. A notebook page by Villard de Honnecourt, depicting some machines and a dove-shaped automaton. [Image from https://upload.wikimedia.org/wikipedia/commons/d/d1/Villard_de_Honnecourt_-_Sketchbook_-_44.jpg.]

Gradually, with a better understanding of the laws of perspective, the drawings of later authors become more and more precise. They were included in "theatres of machines", in specialized books on the fusion of metals, or in sketch books like the famous ones by Mariano di Jacopo Vanni, known as Taccola, (1381–1453), Roberto Valturio (1405–1475), Francesco di Giorgio Martini (1439–1502), Leonardo da Vinci (1452–1519), Vannuccio Biringuccio (1480–1539), Georg Bauer, called Agricola (1494–1555), Agostino Ramelli (1531–1608), Jacques Besson (1540–1573), and Vittorio Zonca (1568–1602), just to mention the most

130 The beginning of scientific technology

well-known. Over time, they became more detailed and understandable. However, their purpose did not change and they were certainly no closer to being what we would call a technical drawing, let alone a construction drawing. An example of this is the mill with a hydraulic wheel with vertical axis by Agostino Ramelli, shown in Figure 7.2.

Figure 7.2. Mill with hydraulic wheel with vertical axis, from *Different and clever machines* by Agostino Ramelli. [Image from https://upload.wikimedia.org/wikipedia/commons/4/47/Cut-away_view_of_water-powered_mill_showing_horizontal_wheel_with_paddles%2C_cog_and_shafts%2C_and_housing_LCCN2006680147.tif]

These books contained a little bit of everything, often ordered in an arbitrary way. They included machines that had evidently been built and used for centuries, together with imaginary machines which could never have worked such as the many perpetual motion machines and construction machines, as well as military machines and automata, and other 'wonders', intended for parties and the entertainment of the powerful of the time.

7.3 OVERTURNING THE STEREOTYPE OF THE 'VILE MECHANICIST'

The "theaters of machines" were a very fashionable literary genre that demonstrated an actual interest in machinery and mechanics. A symbol of this passion was the lathe, the oldest of the machine tools. This was conceived as a 'universal

7.3 Overturning the stereotype of the 'vile mechanicist'

machine tool' in all its versions: first the wood lathe, then the metal lathe and in particular the watchmaker's lathe, the first 'precision machine'.

From the end of the Middle Ages and up to the seventeenth century, the creation of machine parts on the lathe became the hobby of the wealthy bourgeois and nobles.

In the introduction to his *L'art du torneur mechanicien* in 1775, M. Hulot referred to the artisans, noting: "*The art of the turner, in particular, without a doubt, is the one that contributes more to the perfection of the others: there is almost none which does not need its help, I can also assure you that there are many for which it is needed*" **[2]**.

It was not only (and perhaps not even mainly) the professional craftsmen who took advantage of the use of machine tools. The use of the lathe was convenient "*to the people of the church, to exercise and spend part of their time in their properties, avoiding idleness. The nobility can spend the winter days and the rainy days, whether in the city or in the countryside, to have a few hours of leisure time. Loners can have an honest occupation and not get bored in their solitude*" **[3]**.

At the beginning of the nineteenth century, L. E. Bergeron even found it therapeutic: "*The exercise of the lathe is the healthiest among those that reason and doctors prescribe to repair, thanks to moderate movements, to the exhaustion due to lack of activity, to problems due to long convalescence or intellectual work*" **[4]**.

For those who really wanted to be trendy, the instrument of choice was the watchmaker's lathe. Referring to the fifteenth century, Carlo Cipolla wrote: "*In the Renaissance, while clocks became more and more fashionable among the wealthy classes, the watch as a machine attracted progressively the intellectual curiosity of men of culture [...] The paroxysm of interest in watchmaking by intellectuals, professionals or amateurs [...] was reached in the 1600s when the scientific revolution exploded in all its strength*" **[5]**.

The most famous of these hobbyists was none other than emperor Charles V, who was tormented by transience, by *tempus fugit*, from an early age and developed something of obsession with clocks. In addition to collecting clocks and automata, he spent his free time disassembling and reassembling mechanisms. In 1556, when he abdicated and went to 'seclusion' at the monastery of Yuste in the mountains of Extremadura, among the people he took with him, in addition to surgeons, bakers, brewers and butchers, was a master watchmaker called Juanelo Turriano, or Janello Torriani, from Cremona. He was a brilliant engineer, blacksmith and mathematician whose talent earned him the nickname of *New Archimedes*.

Torriano created and restored clocks for the emperor, but he also invented a bit of everything: from a micro-clock placed inside a ring that stung the sovereign's finger at the stroke of the hours; to hydraulic birds that chirped and moved their wings; to mechanical soldiers fighting in battle. Even after the death of his protector, he created a famous female automaton that played and danced; portable mills "so small that you can hide them in a sleeve" but still able to grind nine kilos of wheat a day; and finally the famous *Artificio de Juanelo*, a system of hydraulic machines that transported 40 thousand liters of water a day from the river Tagus to the Alcázar palace, placed a hundred meters higher at the top of the hill in the former capital Toledo. Cervantes, Lope de Vega, Góngora, Quevedo, Baltasar Gracián, all the writers of the Golden Age, mention the ingenuity of Juanelo.

The fact that the bourgeoisie and noblemen were passionate about mechanics had an important side effect. While the artisans had limited economic resources and did not constitute a market capable of motivating machine tool manufacturers to innovate, the rich upper-class hobbyists financed a technological development which led to the creation of machines that facilitated the construction of other machines. While at first glance this might seem a frivolous and useless fashion, it was because nobles such as Louis XVI played with machine tools, together with military advances due to the requirements of artillery, that the development of machines progressed. These developments were fundamental to the later construction of steam engines and thus made the industrial revolution possible.

If the automata were toys for having fun at court and clocks were just instruments to measure time, in the sixteenth century, during the following century they became the symbols of the mechanistic universe of enlightenment in which the human being was reduced to a machine and, freeing himself of the soul, sought to emancipate himself from every servitude, including the metaphysical.

God was no longer perceived as the *Prime Mover* that gave life and movement to the universe, but was instead the *Great Watchmaker*, a cold creator who set in motion an absolutely deterministic mechanism. The craftsman was unconsciously elevated to a symbol of divinity itself, or perhaps we should say that God was lowered to the rank of demiurge (craftsman).

7.4 THE BIRTH OF MODERN SCIENCE

By the second half of the sixteenth century, when Galileo (1564–1642) was active, science had undergone a considerable systematization and was notably different to Greek natural philosophy. The experimental method was establishing itself as the only basis for the verification of theories, although it is not clear whether Galileo's experiments were actually materially carried out or were mere conceptual experiments. Regardless, the instrumentation introduced by Galileo, in particular the telescope, profoundly revolutionized the various disciplines. The use of these instruments was not without its reservations, however. Up to this point, observation had meant viewing with the naked eye, and it remained unclear for some time how much a scientist could also trust what he was seeing through an instrument such as a telescope.

7.4 The birth of modern science

Galileo's contributions to science were many, but was more a systematization and conclusion of an evolving process that had begun in the past than a true revolution. Perhaps the fundamental point was the intensive use of mathematics in the physical sciences or, as we would say today, the large-scale use of mathematical models. That the trajectory of a projectile was parabolic was a discovery that could only be made using a mathematical model whose solution was the equation of a parabola, and similar considerations may apply to many other Galilean discoveries.

However, more than the individual discoveries, what mattered most was the method. We are now able to isolate fundamental aspects of a problem so that we can reduce even very complex problems down to ones that are sufficiently simple to write a mathematical model that can be solved. For example, in the case of the motion of projectiles, the trajectory is parabolic only if the projectile is presumed to be a material point, the weight force is directed in a single direction (that is, the motion occurs in a sufficiently small area of the Earth's surface to make it possible to conflate the spherical surface with its tangent plane), and the aerodynamic forces are negligible compared to the weight. The problem is then reduced to a simpler one that is solved 'in closed form', by obtaining a solution expressed with an algebraic equation, which in this case yields the trajectory.

A reductionist approach of this kind is not substantially different from what Archimedes did to demonstrate the principle that bears his name, but that came from the time of Hellenistic science and was not utilized so clearly. Following Galileo, the principle became a constant of science, in part because there were no viable alternatives until the introduction of computers in recent decades. We will return later to the subject of criticism of reductionism, which often means a criticism of modern science in general.

Another very important aspect of the Galilean approach was the use of science to solve practical problems of a technological nature. Behind many of Galileo's studies there were concerns as to how they could be applied, but even here Galileo followed a theme dating back to the Hellenistic period, one which had never truly been addressed but now took on much greater importance and was bound to grow in the following centuries. The more science progressed, the more it became applicable, but the more it needed sophisticated investigative tools and therefore a more advanced technology to be able to progress further. This remains true today.

Benjamin Farrington, in his essay dedicated to Francis Bacon (1561–1626), a contemporary of Galileo and a proponent of the experimental method, summarized the idea that permeated the life and work of the philosopher with the words: *"That knowledge should bring its fruits in practice, that science should be applicable to industry, that men had the sacred duty to organize themselves to improve and transform the conditions of life"* **[6]**.

Even in the fields of research that could be considered the furthest away from having an application, such as astronomy, Galileo sought an applicative outcome.

After discovering the four main satellites of Jupiter, he had the idea of compiling tables in order to obtain the time of the fundamental meridian of Earth by using observations of Jupiter, in order to obtain the longitude of the observation point.

Evangelista Torricelli (1608–1647), a pupil of Galileo, invented the mercury barometer, measured atmospheric pressure and created the Torricellian vacuum, putting an end to centuries of discussions on the subject. Although similar results had been obtained in the Hellenistic era, Torricelli's work laid the foundations for the steam engine, at least in its original form as an atmospheric machine.

7.5 THE STEAM ENGINE

As has already been noted, the creation of a thermal engine was attempted by Heron of Alexandria, but his aeolipile was little more than a scientific toy and had no practical applications. In fact, the thermal engine probably approached what was the limit of pre-scientific technology.

The first practically usable steam engine was the machine created by Thomas Savery (1650–1715), patented in 1698. This was a pump suitable for pumping water from mines, a device that was particularly relevant to dealing with the very serious problem of flooding afflicting the Scottish mines. The principle was very simple. A sealed container was connected alternately through two valves to the suction inlet at the bottom of the mine or to the outlet at ground level. The depth from which the machine could remove water was limited to less than about ten meters, due to atmospheric pressure.

The vessel was connected to a boiler and filled with steam. Then the steam supply was closed and the container cooled, allowing the steam to condense thus making a vacuum. When the suction valve was opened, the water was drawn from the bottom of the mine and filled the container. Then the intake valve was closed, the exhaust valve opened and the vessel refilled with steam, which pushed the water away and discharged it to the highest level. The higher the steam pressure, the higher the level at which the water could be discharged, although at that time it was preferred not to use pressures much higher than atmospheric pressure for safety reasons. The cycle was then repeated, removing a certain amount of water each time.

It is clear that a machine of this kind could only be conceived by someone who was aware of the existence of atmospheric pressure and could produce a vacuum. These principles became better understood following the experiments of Torricelli and Otto Von Guericke.

The major limitation of the Savery machine was its inefficiency, and therefore its high fuel consumption. At every cycle, it was necessary to reheat the container in which the steam was expanding and then to cool it during condensation.

7.5 The steam engine

However, at the time of Savery's machine, the concepts of energy and power had not yet been defined, nor had the concept of efficiency, so no-one would have known how to increase the latter.

After Savery, Newcomen built a new engine, again for use in pumping water from the mines. This was the first machine in which the steam, acting inside a cylinder, moved a piston and therefore the thermal energy produced by combustion was transformed into mechanical energy (Fig. 7.3). However, it was still an atmospheric machine, with the steam used to create a vacuum (obviously a partial vacuum) in the cylinder, allowing atmospheric pressure to move the piston downwards. What actually performed the useful work was therefore not the steam pressure, but the atmospheric pressure. The piston moved a rocker arm, which in turn drove a rod connected to the pump located at the bottom of the mine shaft.

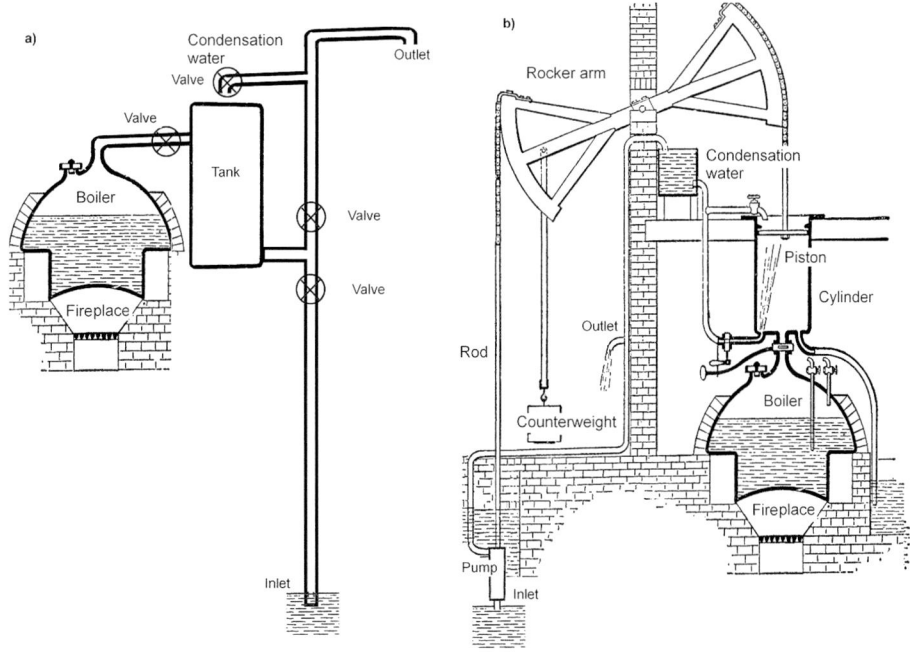

Figure 7.3. Schematics of the Savery (a) and Newcomen (b) steam engines.

The condensation of steam still occurred in the cylinder, but in this case was assisted by a jet of cold water coming from either a secondary tank supplied by a small auxiliary pump and driven by a second rod, or by the main pump itself. It should be noted that the machine was manually operated by means of valves, which connected the cylinder to the boiler or to the condensation water tank.

Use of the engine quickly spread and Newcomen, who had associated himself with Savery, produced a hundred of them, yet its performance was not substantially better than that of Savery's engine.

In addition to pumping water from mines (coal mines, but also tin mines), these steam engines could be used to operate machinery. In this case, the water was pumped to a higher level and its potential energy was utilized through a normal water mill.

To progress further, however, required further science, which was provided in this case in the person of James Watt (1736–1819). After attending the University of Glasgow and becoming a friend of several professors – particularly Joseph Black (1728–1799), the scientist who introduced the concept of latent heat – Watt opened a workshop at the university with three of the professors. While repairing a model of a Newcomen engine on behalf of the university, Watt thought he had found the cause of the low efficiency of the engine. While condensing the steam in the cylinder, much of the heat was lost in also cooling the cylinder itself, which then had to be heated again during next cycle. Watt suggested introducing a separate condenser.

Later, he moved to Birmingham, where he started a company with Matthew Boulton and began to build steam engines that became commercially successful. In fact, the separate condenser greatly increased the efficiency of the engine and reduced the fuel consumption by at least a factor of three. Moreover, by using a planetary gear and then the simpler crank-connecting-rod system, he transformed the rocking motion of the rocker arm into a continuous rotary motion of the output shaft. With the introduction of a speed regulator and other improvements, the steam engine transformed from a water pump for mines into a general-purpose, safe, reliable and relatively inexpensive machine.

Watt's familiarity with the scientific environment of his time is verified by his election to the Royal Society in 1785 and later to the Paris Academy of Sciences as a foreign member. Many scientists studied the steam engine, developing the discipline that is now called thermodynamics, including Sadi Carnot (1796–1832), who succeeded in creating a theoretical model of an ideal thermal machine; Rudolph Clausius (1822–1888), who formulated the second principle of thermodynamics and introduced the concept of entropy; and William Thomson, Lord Kelvin (1824–1907), to name but a few.

The steam engine is most closely associated with the Industrial Revolution, although to begin with the industry relied more on hydraulic and wind power than anything else. At some point, however, the machines of the Industrial Revolution were increasingly powered by the energy of coal and later by other fossil fuels. The steam engine then became essential in transportation, with steam railways and steam navigation.

7.6 THE DREAM OF FLYING

We have already mentioned the passage by Roger Bacon in which he stated that, in the future, flying machines would be built. It was obviously an unattainable dream at that time, yet some people thought they could do it. Drawings of flying machines can be found in some of the "theatres of machines", and particularly in the notebooks by Leonardo da Vinci, while some attempts at flying were performed by would-be-followers of Daedalus and Icarus.

These attempts had very little scientific basis before Torricelli measured the value of atmospheric pressure and von Guericke built his vacuum pump. At that point, it was possible to conceive of making a vacuum in a container, which could fly by exploiting Archimedes' thrust due to the displaced air.

At least, this was the idea forwarded by the Jesuit Father Francesco Lana de Terzi (1631–1687), a mathematician and naturalist, who drew a sketch of a flying ship hanging from four hollow copper spheres in which a vacuum had been created. His idea was also unfeasible – a copper sphere thick enough to withstand atmospheric pressure would have been much heavier than the aerostatic thrust could have supported – but at least it was based on a reasonable theory. It could never work with vacuum, but if the spheres could be filled using a gas lighter than air – hot air or hydrogen – and were just simple bags of light canvas or paper, the theory could become viable.

The first solution was realized about a century after Father Lana, by the Montgolfier brothers, Joseph Michel (1740–1810) and Jacques Étienne (1745–1799). Both had a scientific background and worked in their father's paper mill on technical tasks. However, they did not understand that the hot air balloon worked because of the expansion of the air which therefore became lighter. Instead, they believed that smoke contained a particularly light gas (then called Montgolfier's gas). Initially it was feared that air was not breathable at the altitude reached by the balloon, so they were not allowed to fly a balloon with human beings until it was proven that animals were able to survive the flight. In 1783, the first aeronauts were a sheep, a goose and a rooster, in the same way that the first astronauts were dogs and chimpanzees just a couple of centuries later. That same year, the first men to fly were Pilâtre de Rozier and the Marquis d'Arlandes and once they had proven that flying had no negative consequences on health, they were followed by many others. Balloon flying soon became fashionable.

Interestingly, the hot air balloon could certainly have been made in a pre-scientific environment. It is said that what gave Joseph Montgolfier the idea was his observation of some clothes, which tended to billow upwards while put over a bonfire to dry. Indeed, it is perhaps surprising that hot air balloons had not been invented earlier, for instance by the Chinese, perhaps as a by-product of the invention of paper. The fact that the Montgolfier brothers worked in the paper industry is significant. The *Montgolfières*, as hot air balloons were immediately called, were constructed using a layer of sackcloth, airproofed inside with three layers of light paper and reinforced on the outside by a fishing net.

In contrast to the hot air balloon, its competitor the hydrogen balloon could not have been built in a pre-scientific world. Hydrogen was discovered by the English chemist and physicist Henry Cavendish (1731–1810) by adding sulfuric acid to scrap iron, tin or zinc. Given the lightness of hydrogen, the physicist Jacques Alexandre César Charles (1746–1823) thought of using it to make a hydrogen balloon.

The difficulties in doing so were significant, mainly because it was almost impossible to obtain the gas in sufficient quantities to fill a balloon, but also because it was extremely difficult to contain the gas, which tended to escape through the smallest pores of the casing. To resolve this, Charles, together with the Robert brothers, developed a process to airproof a light silk cloth with rubber dissolved in turpentine. The gas was produced by pouring about 250 kg of sulfuric acid onto almost half a ton of scrap iron and was then transferred to the balloon through lead pipes.

A balloon capable of lifting about 9 kg from the ground (including the casing) was built through public funding and was launched from the Champ de Mars – where the Eiffel Tower is now – in 1783, the same year in which the Montgolfier brothers launched their hot-air balloon. The flight was successful, and a bigger balloon capable of carrying two people was built. Charles and Nicolas-Louis Robert flew for 36 km in just over two hours, only ten days after Pilâtre de Rozier and the Marquis d'Arlandes flew on the Montgolfier brothers' balloon. After landing, Charles took off again, reaching an altitude of 3000 m and showing that the hydrogen balloon (which was immediately called *Charlière*, in his honor) was much more controllable than the hot air balloon.

The following year, an elongated dirigible was made, but attempts to control the motion of the balloon using oars proved to be unsuccessful. In order to develop a real airship, a suitable thermal engine needed to be developed. Charles subsequently used aerostats for scientific research and invented, or perfected, instruments useful for aerostatic navigation.

The technologies of both hot-air and hydrogen balloons developed, but the latter took over until the Hindenburg disaster of 1937, when hydrogen was replaced by helium. Hot air balloons found a new era of popularity starting in the 1960s, thanks to the use of nylon for the casing and propane to heat the air. All the most important aerostatic enterprises of the eighteenth and nineteenth centuries (such as the flight across the Channel in 1785) were carried out using hydrogen balloons.

The news of the first flight of an "air ship" quickly spread around the world, arousing wonder and enthusiasm. In Italy, the poet Vincenzo Monti wrote the *Ode al Signor di Montgolfier*, a poem celebrating progress and science in which

Montgolfier is compared to Jason, the mythical inventor of the art of navigation. In the poem, the author also talked about the hydrogen balloon, chemistry and the discovery of hydrogen, as if Monti had realized that in reality the hot air balloon, unlike the hydrogen balloon, was not a result of science, which his poem was intended to celebrate. The poem ended with the prospect that science could in the future also defeat death: "*break/ even the bolt of death/*", and make man similar to the gods.

Throughout the nineteenth century, flying "like birds", that is, flying using a heavier-than-air aircraft, remained an elusive dream, a feat which was apparently more difficult than aerostatic flight and, in the opinion of many scientists, was actually impossible. During this period, there were plenty of 'scientific' demonstrations of its impossibility.

In fact, building an aircraft required greater knowledge of the field of aerodynamics, to evaluate the forces acting on a body moving in a fluid. This problem was studied intensely by Jean-Baptiste Le Rond, called d'Alembert (1717–1783), who tackled the problem, as is usual in the scientific field, by trying to isolate the key aspects and make the appropriate simplifications. Assuming that the fluid was incompressible (a correct assumption in the subsonic field), not viscous, free from vortices and in stationary condition (all reasonable hypotheses), he reached a conclusion known as the d'Alembert Paradox, which he expressed with these words: "*I do not see how, I admit, it is possible to explain the resistance of the fluids by means of the theory in a satisfactory manner. It seems to me, on the contrary, that this theory, treated and studied with deep attention, gives, at least in most cases, absolutely no resistance, a singular paradox that I leave to the scholars to explain*" [7]. This applied not only to resistance, but in general for any aerodynamic force.

D'Alembert's paradox had major consequences for the new science of fluid dynamics and caused a dichotomy between the theoretical and the experimental approach. The theory showed that something heavier than air could never fly, which meant it was necessary to resort to experimentation. The first experimental approach was to create aircraft models and test them in free flight, or by pulling them like kites. Another possibility was to use a rotating arm with the model of an aircraft secured to its end, to measure the forces that the latter exerted on the arm once in motion. However, the best solution was to build a wind tunnel, a channel able to blow a current of air over a stationary object to simulate the motion of the object in still air.

The most important aerodynamicist of the first half of the nineteenth century was Sir George Cayley (1773–1857), a parliamentarian, scientist and rector for many years at the Royal Polytechnic Institution. In 1799, after studying the flight of birds for years, Cayley defined what would later become the configuration of the airplane. He postulated that the three fundamental functions – the generation of lift, propulsion and stability – which in birds are performed by the same parts of their body, should be entrusted to three distinct organs (the fixed wing, the engine and the tail planes, respectively). On closer inspection, this was a radical application of the reductionist approach.

Following this approach, he apparently decided that the problem of propulsion could be postponed until engines with a sufficient power/weight ratio would be available and instead focused on the construction of gliders. He started on a small scale (1804), then built one large enough to carry a child, before finally constructing one of a larger size in which he had his coachman, who was not at all enthusiastic about becoming a pilot, fly at the controls.

In 1809, Cayley began the publication of a three-part treatise *On Aerial Navigation*, the first rigorous treatise on the subject. In 1866, he founded the Aeronautical Society of Great Britain and two years later he organized the first aeronautical exhibition in London.

Following Cayley, other pioneers addressed the problem of flying and patented various innovations that gradually brought the solution to the problem closer. In 1871, Francis Herbert Wenham and John Browning built the first wind tunnel, which was soon imitated by other scholars in Europe and the United States.

Other pioneers experimented with gliders, sometimes pulled by animals. These attempts were often very successful, such as those by Otto Lilienthal (1848–1896), Octave Chanute (1832–1910) and Samuel Pierpont Langley (1834–1906). The former built and tested several gliders, completing more than 2000 gliding flights, some of which reached distances of about 250 m. He left abundant documentation, including 145 photographs, which heavily influenced both other experimenters and public opinion. He died in a flight accident before he had the chance to install a steam engine, which he had patented a few years earlier, onto a glider to transform it into a powered airplane.

Chanute was a civil engineer and urban planner specializing in the design of iron bridges. In 1890, he retired as a civil engineer and began to work with flying machines again, this time practically full-time. In 1894, he published *Progress in Flying Machines,* a book that deeply influenced all those who tried to build a heavier-than-air vehicle that could fly. At the end of the nineteenth century, Chanute filled the role that Cayley had played at the beginning.

The configuration he identified was that of a biplane, with rectangular wings and fairly high aspect ratio. This was quite different to the sometimes-extravagant configurations in the shape of bird or bat wings adopted by many other pioneers. Most airplanes followed similar configurations until after the end of World War I.

His structural expertise played a key role in the construction of the first airplanes, a skill that Chanute shared with anyone interested in the subject, writing articles and books, exchanging letters with them and even visiting them during tests.

The last of the pioneers, Samuel Pierpont Langley (1834–1906), was certainly the most learned (in the academic sense) but also the unluckiest of the three.

7.6 The dream of flying

A famous astronomer and secretary of the Smithsonian Institution, he conducted rigorous experiments in aerodynamics at the University of Pittsburgh. In 1891, he published *Experiments in Aerodynamics*. In 1896, his Aerodrome No. 5 made two powered flights of 1,005 m and 700 m at a speed of about 40 km/h. In the same year, the subsequent Aerodrome No. 6 flew for 1,460 m. They were what we would call today free flight model aircraft, even though they were large. Following these successes, he received a $50,000 contract from the government to develop a human-carrying airplane. In 1901, Langley successfully flew a one-quarter scale model of the aircraft and turned his focus to the full-scale airplane. After several difficulties, his assistant managed to achieve 39 kW from his water-cooled five-cylinder radial engine, a very high power for the time.

At that point, he built the full-scale airplane and installed the engine, but when it was tested in 1903, the Aerodrome proved not to be strong enough, only marginally stable and difficult to control, falling twice into the Potomac River off the floating platform from which it had launched. Only nine days after the second attempt, the Wright brothers succeeded in making the first powered flight and Langley desisted from further attempts.

Some have argued, even recently, that in those same years (1901–1902) Gustave Whitehead probably made several flights with one of his powered aircraft, but most historians believe that those flights never happened.

When Orville (1867–1912) and Wilbur (1871–1948) Wright began their experiments at the end of the nineteenth century, they started by collecting a lot of documentation. They even wrote to the Smithsonian, asking the prestigious institution to provide them with an extensive bibliography on flight. Once they received the information, they read everything they could find on the subject. Their study of what others had already done convinced them that the first two points identified by Cayley, propulsion and lifting force, had been solved sufficiently, so they focused on the third, control. Here, they introduced a novelty. Instead of trying to provide 'automatic' stability to their aircraft, they preferred to build an intrinsically unstable aircraft and gave the pilot the role of stabilizing the machine. In this, they were perhaps influenced by their experience with bicycles, and this was probably the key to their success.

Before starting with flight testing, they asked Chanute for advice on the best place to do so. He advised them to move to the Atlantic coast for the constant wind and the sandy ground, which would minimize the consequences of any accident. The same advice was received from the U.S. Weather Bureau, so they chose the town of Kitty Hawk, in North Carolina.

The 1900 test campaign was first carried out using a biplane glider, of the type used by Chanute, with whom they were always in close contact. After the partial success of that test campaign, they built another, bigger glider for the following year. They were somewhat disappointed with the performance of the second prototype, and began to doubt the values of the aerodynamic coefficients

142   **The beginning of scientific technology**

which had been measured by Lilienthal. To check this, they built a small wind tunnel, which led to an impressive series of experiments on wing sections and aircraft models.

Following this experimentation, they designed an even bigger third Flyer, which they tested in 1902, completing hundreds of flights with the main purpose of learning to fly it precisely. Not only did the glider fly, but it was completely controllable. At the end of 1902, the two brothers had a working aircraft that could fly in a stable and controlled way. They were the only people who knew how to fly an airplane.

The time had come to add the engine. Unable to find a light but powerful enough engine on the market, they designed one that was built by one of their assistants. Having also failed to find in the literature a theory that would allow them to design the propellers, they experimented again in the wind tunnel, developing for the first time a powerful and light power unit.

In the 1903 test campaign, they met with some problems related to the engine and a small accident caused minor damage to the aircraft. On December 17, after repairing the aircraft, Orville Wright managed to take off and fly for 37 meters. The historic flight is depicted in a famous picture (Fig. 7.4). That same day they flew three more times, alternating at the controls, with Wilbur finally achieving 260 m in 59 seconds.

Figure 7.4. The first flight of an airplane (December 17, 1903). [Image from https://upload.wikimedia.org/wikipedia/commons/e/e3/First_flight3.jpg.]

Over the following two years, the Wright brothers continued to improve their airplane, building a second and then a third version with which Wilbur flew almost 40 km in 39 minutes. By then, flying in a heavier-than-air vehicle was a reality.

Nevertheless, d'Alembert's Paradox remained unexplained. The airplane flew, behaving as the wind tunnel experiments had predicted, but the theory still failed to explain how this was possible. It was only in 1904 that Ludwig Prandtl (1875–1953) provided a theoretical explanation, which was immediately accepted and is still reported in all the texts on aerodynamics. The fracture between theoretical and experimental aerodynamics was thus healed, even though aerodynamics would remain an eminently experimental science for a long time.

7.7 ELECTROMAGNETISM

Electromagnetic phenomena have been known for millennia, at least as far as static electricity is concerned. The Greeks, for example, knew very well that a piece of amber, if rubbed, was able to attract small fragments of many materials and that a magnet attracted pieces of iron. The former observation is attributed to Thales, while Plato spoke of the properties of the magnet in his *Timaeus*. The term electricity comes from *electron*, which means amber in Greek.

Although it does not seem that Hellenistic science dealt with electromagnetic phenomena, both Pliny the Elder and Lucius Anneo Seneca spoke about lightning, although they probably did not connect it to static electricity.

Girolamo Cardano (1501–1576) was perhaps the first to distinguish between electric and magnetic phenomena. At the end of 1600, the first electrostatic machines were built, such as that by Otto von Guericke (1602–1666). Given the spectacular nature of electrostatic phenomena – sparks, diffuse brightness, crackles – electricity was a curiosity that attracted widespread attention in the eighteenth century gatherings of intellectuals. However, finding useful applications for it would require the development of a theory that would allow scientists to understand the phenomena and to reproduce them on a larger scale.

Just as it was thought that heat was due to the movement of a fluid, called the caloric, so was electricity seemingly linked to a particular electric fluid. Benjamin Franklin (1706–1790), the inventor of the lightning rod, established the relationship between lightning and electricity and explained the functioning of the Leyden bottle, the first capacitor, which had been invented in the 1740s. His explanation, though still based on the electric fluid, laid the foundations from which the scientists who followed him, such as Michael Faraday (1791–1867), Luigi Galvani (1737–1798), Alessandro Volta (1745–1827), André-Marie Ampère (1775–1836), Georg Simon Ohm (1789–1854) and Hans Christian Ørsted (1777–1851), were able to explain electric and magnetic phenomena.

Unlike the situation with aerodynamics, where the theory and the experimental results remained in contrast for more than a century, for electricity the theory developed by James Clerk Maxwell (1831–1879), and summarized by the equations that even today carry his name, perfectly explained the results obtained in the laboratory.

Around 1830, Faraday created the first generator able to transform mechanical energy into electrical energy. From this, electrical energy was made available in such quantities that electricity was no longer a curiosity but something that could be used in practice. In a short time, it became possible to transmit electric energy and to transform it back into mechanical energy using electric motors.

Initially, the steam engine schematic was copied, where an electromagnet attracted an anchor that operated a rocker arm like the piston of a steam engine. In turn, the rocker made the crankshaft rotate via a crank-connecting-rod system. It was not long before the components with a reciprocating motion were eliminated and left the electric motor with only rotating elements.

Various types of generators and motors were developed before the end of the nineteenth century, working with direct current and alternating current. Transformers were developed for the latter and it became possible to transport electricity over long distances.

In a few short years it was realized that electricity could do more than just transform mechanical energy into electricity, transport it and then turn it back into mechanical energy. Starting from a huge generating machine, the energy could be distributed to several users, who could operate a large number of small machines, each with its own small motor, or even use multiple motors to drive the various 'axes' of a single machine.

Such uses of the energy were soon accompanied by other specific uses, developed by many inventors. Electricity could be used to supply light (the invention of the light bulb by Thomas Alva Edison, 1876), to carry information (first the telegraph by Samuel Morse, 1844; then the telephone, by Antonio Meucci and then Alexander Graham Bell, 1876; and finally the radio by Guglielmo Marconi, 1894), to record sounds (the phonograph, by Edison, 1877) and other uses. Many inventions were controversial, and the names and dates given above are only indicative. Some inventors produced a large number of inventions, such as Edison, who obtained more than 1,000 patents.

The applications for electricity were all derived from the theoretical developments that the group of scientists mentioned above, and others, perfected in less than a century, and which culminated and were systematized into Maxwell's equations. It is probable that it could only have been done that way: it was not possible to develop applications for something as intangible and invisible as electricity without developing a full understanding of electromagnetism as a science.

Even so, many provided interpretations of electric phenomena that bordered on magic and esotericism. After Galvani's experiments, which made the muscles of dead frogs contract by applying electrodes – in practice, making dead bodies move – many thought that electricity was a vital fluid, something magical, and animal electricity theories spread. A symptom of this trend was the novel *Frankenstein* by Mary Shelley, published for the first time between 1817 and 1818.

In the last quarter of the eighteenth century, the theory of animal magnetism had also developed. Its author, Franz Anton Mesmer, a physician, alchemist, astrologer and esotericist, claimed to be able to heal many diseases using a so-called magnetic fluid. His doctrine, called mesmerism, was repeatedly examined by scientific commissions that mostly stated that the healings took place due to self-suggestion. Nonetheless, there were also scientists who believed in Mesmerism.

7.8 NUCLEAR ENERGY

Nuclear technologies, even more than electrical ones, could only develop as a result of modern science. Indeed, it can be asserted that it would never have been possible to develop them in the context of a non-scientific technology. They are a product of the twentieth century and do not have forerunners. Conceptually, one of the results of atomic physics is the equivalence between energy and matter. From a quantitative point of view, it was realized that enormous quantities of energy can be obtained from transforming a very small amount of matter.

Nevertheless, the pioneers of nuclear physics and those who discovered the matter-energy equivalence, such as Niels Bohr and Albert Einstein, believed that this possibility could not be exploited from a practical point of view. In 1932, Ernest Rutherford discovered that large amounts of energy are released when a lithium atom is split by a proton into a particle accelerator. However, he also defined as "moonshine" (nonsense) the prospect of exploiting this behavior for practical purposes.

In the 1930s, Enrico Fermi performed many experiments using slow neutrons with which he bombarded uranium nuclei to create nuclei of transuranic elements. In 1938, a group of German physicists, including Otto Hahn and Fritz Strassmann, replicated these experiments and succeeded in splitting the nucleus of the atom of uranium into two almost equal parts, with the release of huge amounts of energy. The process was called fission and it was discovered to cause the emission of neutrons, meaning that it could therefore be exploited to produce a chain reaction, able to sustain itself.

After the confirmation of this phenomenon by Frédéric Joliot-Curie in 1939, it was realized that a process capable of delivering such large quantities of energy had potential military applications as explosives, which at the time was more interesting for many governments than its possible civilian application as an energy source. In the United States, Great Britain, France, Germany and the Soviet Union, many scientists tried to make their governments aware of both the military possibilities of the new technology and the danger that such technology could be developed by the enemy. In particular, the Allies were very worried about the possibility that such a weapon could be built by Nazi Germany.

In 1942, the Chicago Pile-1, the first reactor built by the Fermi group, reached critical condition and provided a fundamental contribution to the realization of the first atomic weapon. Military technology will be widely discussed in Chapter 9. Civilian uses of nuclear technology started to be seriously considered in 1945, just after the end of World War II, with the publication of a document entitled *The Atomic Age*. Unlike many military studies, this was widely distributed, and it suggested a scenario in which fossil fuels would be completely substituted by nuclear energy. Nobel laureate Glenn Seaborg wrote that *"there will be lunar spacecraft propelled by nuclear energy, nuclear artificial hearts, plutonium-heated pools..."*, listing a wide range of applications.

Nevertheless, since the beginning, the focus has been on just two applications: large power plants and ship propulsion systems. In 1951, electricity was generated by a nuclear source for the first time, while in 1953 the first nuclear reactor for ship propulsion was developed. The first nuclear submarine, the USS Nautilus, was launched in 1954.

In his speech *Atoms for Peace* at the United Nations in 1953, U.S. President Dwight D. Eisenhower launched a policy of liberalization of nuclear technologies for peaceful use and encouraged private individuals to design and build nuclear power plants. It seemed that nuclear power had opened up an era of abundant energy at low cost for everybody, even though the goal at that time remained lowering the costs so that the new technology was competitive with conventional ones. The United States, the Soviet Union, Britain, France and Italy became the leading countries for this new technology and built many nuclear power plants around the world.

With regard to ship propulsion, many countries built nuclear submarines, and the United States also built nuclear aircraft carriers. The first commercial nuclear ship, the American N/S Savannah (Fig. 7.5), was launched in 1959 with the aim of demonstrating the possibility of building a nuclear-powered cargo ship and supporting the American program *Atoms for Peace*. However, in 1972, after years of service without the slightest accident, it was withdrawn from active service to save on operating costs, which were higher than those of a similar conventional ship. It would only have become competitive a few years later, due to the increase in the cost of oil.

7.8 Nuclear energy

Figure 7.5. The N/S *Savannah*. [Image from https://commons.wikimedia.org/w/index.php?title=Special:Search&limit=20&offset=40&profile=default&search=NS+Savannah&&advancedSearch-current=%7b%7d&ns0=1&ns6=1&ns12=1&ns14=1&ns100=1&ns106=1#/media/File:(The_nuclear_ship_Savannah_passing_under_the_Golden_Gate_Bridge_in_San_Francisco,_California,_on_its_way_to_the..._-_NARA_-_542141.tif.]

Of the four nuclear cargo ships built (besides the Savannah, there was a Japanese, a German and a Russian ship) only the Russian one, the *Sevmorput*, remains in service. In addition to these four cargo ships, ten nuclear icebreakers were built in the Soviet Union and Russia, with others still planned.

Other applications of nuclear energy – apart from a few space applications – have not materialized. In the 1950s and 1960s, it was expected that nuclear energy would be used in many different applications, from automotive to aeronautics, to fixed installations for various types of industries. However, safety and security problems have so far caused several projects of this kind to be discontinued. The nuclear-powered airplane was the one that came closest to construction, both in the United States and in the Soviet Union. In the USA, a B36 bomber was converted and flown with a functioning nuclear reactor, but this was not connected to the engines which still operated using fossil fuels. The program was suspended before it could actually fly using nuclear power. The prevailing idea has always been that nuclear energy is convenient only for large power plants that transform energy into electricity, which is then distributed to drive the various uses. For example, the generation of electricity by nuclear power stations can make the use of electric vehicles feasible, whose batteries could be recharged in a convenient and economical way. The reason this has not happened is linked above all to safety and security. The first – essentially safety against accidents – would be endangered by the proliferation of small nuclear units, perhaps on vehicles, which may be involved in normal traffic accidents but with the potential to contaminate large areas. The second – substantially security against sabotage, terror attacks, theft of nuclear material, and other similar acts – would be compromised by the difficulty of monitoring numerous small plants rather than a few larger power units.

Since the 1960s, there has been a strong and growing opposition from a sector of public opinion against the use of nuclear energy, precisely for safety reasons. Such opposition has led many governments, including those of countries like Italy, which were among the leaders in nuclear technology, to abandon it completely. The opposition, which objectively can be deemed unreasonable since nuclear technology itself is extremely safe, is motivated by accidents which have occurred in the past due to the failure of some operators to comply with a number of safety regulations.

The worst accident, and the one that most influenced public opinion, was that of Chernobyl in 1986. This was an obsolete and intrinsically unsafe reactor, which had already had an accident four years earlier, on which an unauthorized accident simulation had been carried out after the disarming of a number of automatic safety devices. The previous accident had not been communicated to the public and to the control bodies.

The problem with nuclear technology is that it is a further step along the road which began with the introduction of fire and then with gunpowder and many other technologies, a road that permits small or tiny actions – or even errors – to produce consequences of catastrophic dimensions.

Moreover, in the collective imagination, nuclear technologies are connected with nuclear bombs – even if a reactor can never explode like a bomb – and the danger is mainly related to the possible spread of radioactive substances. Such a danger is perceived as particularly frightful, given that radiation is invisible and evades all human senses.

The poor management of some plants, particularly in the former Soviet Union, which led to serious accidents such as Chernobyl, has then made things worse in terms of acceptance by the public. This may lead to the loss of the enormous advantages that nuclear energy can bring.

In fact, nuclear power plants are much less polluting, both in terms of atmospheric pollution because they do not release combustion gases or gases that produce greenhouse effect, and in the pollution related to the extraction of fuels such as oil or coal. This is why countries like China and India have revived the construction of a large number of nuclear power plants as part of their programs to reduce pollution.

Undoubtedly, however, nuclear energy obtained from the fission of heavy elements is only a transition towards solutions which are much more convenient from all points of view: energy from the fusion of light elements. In the military field, the transition from fission to fusion was extremely quick and the first hydrogen bomb was tested in 1952. Nevertheless, after decades of research, it has still not yet been possible to produce a fusion reactor capable of producing more energy than it requires to 'ignite'. However, the advantages of fusion are huge, ranging from the significant reduction in radioactive waste – one of the major problems of fission reactors – to the lower quantity of radiation produced and the availability and low cost of the nuclear fuel (deuterium, a relatively abundant hydrogen isotope found in sea water, rather than uranium or plutonium). Research in this field is very active.

We can hope that it will be implemented within a reasonable timeframe. Recent estimates speak of about fifteen years, but after so many statements of this kind which have proven to be incorrect, it is advisable not to try to forecast the timing.

One of the advantages of nuclear fusion may be the possibility of creating devices of small size, able to feed the final devices directly. The miniaturization of fusion systems undoubtedly has fewer safety problems than fission devices, if only because the elements in a fission reactor which must be constantly monitored, such as uranium or plutonium, are not present.

7.9 THE ÉCOLE POLYTECHNIQUE

While the press, the compass and the gunpowder, the three great inventions which marked the end of the Middle Ages, were the result of pre-scientific technology and had originated in China – even though they could only be developed in the West – the steam engine, the airplane and the devices based on electricity, the three great inventions of the eighteenth and nineteenth centuries, could not, by their very nature, ignore science and were therefore necessarily the prerogative of Western civilization.

Thermal engines certainly had precedents dating back to the Hellenistic times, while aerodynamics had objective difficulties due to the lack of understanding of d'Alembert's Paradox and had to rely on experimentation. In contrast, electromagnetism quickly found a theoretical synthesis in Maxwell's equations and, perhaps for this reason, had a much faster development than the other two.

It took almost two centuries to go from understanding atmospheric pressure to the realization of atmospheric machines, and then to the practical use of steam engines in fixed installations, rail vehicles and ships. It took more than a century to make a heavier-than-air vehicle that was capable of actual flight. For electricity, on the other hand, it took only a few decades to pass from the first studies to the construction of large power plants and the global diffusion of the telegraph.

In less than 50 years, we reached the electrification of the railways and the invention of the radio.

For the use of nuclear energy, the significant invention of the twentieth century, it only required a few years to go from the demonstration (the Chicago Pile-1) to military use and then to power plants. For the realization of nuclear devices, the theoretical knowledge of nuclear physics was essential, and the technology is based on what is perhaps the most abstract and least understandable part of theoretical physics, in terms of empirical reasoning.

If science had to be heavily involved in these technological achievements, then the people who had to put this revolution into practice needed to have very different backgrounds to those of their predecessors. As we have said, the professions of the *architectus* and of the *ingegnarius* began to differentiate from each other beginning in the seventeenth century, and the design and construction of civilian and military machinery became the remit of the latter. While those who wanted to devote themselves to the medical profession, or wanted to learn a science such as mathematics or physics, could enter a university, there were no specific schools for the profession of engineer or architect. Those who wanted to start an engineering career were given a generic, possibly scientific preparation, after which they entered the workshop of an expert engineer from whom they learned the trade. In the mid-eighteenth century, the first attempts to start schools of engineering and architecture met with little success. For example, a school for engineers in Saxony closed after a few years of trying.

In 1747, the *École des Ponts et Chaussées* was founded in Paris (the current name dates back to 1775). It is still there, and today is the oldest school for engineers in the world. Its task was to create engineers to be employed by the public administration for the construction and maintenance of bridges and roads.

In 1794, also in Paris, the *École Polytechnique* was founded. For more than a century, this was the model that inspired the schools for engineers in Europe and in the world, and from which the word 'Polytechnic' is derived (even though in the English-speaking world the name Polytechnic is still given to non-university lower level institutions). In 1804, the École was militarized by Napoleon, who made it a military school of the highest level.

The traditional formula of the École Polytechnique is based on a two-year preparatory period in which students learn the scientific fundamentals, in particular mathematics, physics and chemistry, followed by a three-year application period in which students continue with engineering studies. From these names, it is clear that technology is understood to be an application of science. At the end of the eighteenth century, this showed that the transition from empirical technology to scientific technology was now complete, at least in the academic environment.

The engineer had to be skilled in the various scientific disciplines, to a level not much below that of a specialist in the respective sciences. Only after having acquired the fundamental scientific knowledge could they proceed with their application studies.

However, it should be noted that the engineers who came out of these schools were mainly intended for the army and the public administration (at least in France, where the state was very centralized and could count on an administration of very high quality), while those carrying out technical tasks in industries were often self-taught or had a less formalized technical culture.

The professors of the École Polytechnique were often scientists of the level of Gaspard Monge (1746–1818), who 'invented' projective geometry, the basis of technical drawing; Lazare Carnot (1753–1823), a politician but also a researcher of geometry and mechanics, as well as his son Sadi, a thermodynamics expert; Joseph-Louis Lagrange (1736–1813), a mathematician and one of the initiators of analytical mechanics; and Claude-Louis Berthollet (1748–1822), a chemist. The last two had studied at the University of Turin, in Italy.

As we have said, the strongly theoretical and scientific approach of engineering studies, according to the French school's doctrine, remained in continental Europe until at least the last half of the twentieth century. It was accompanied by a strong unity among engineering that was free of excessive specialization, unlike in the English-speaking world where that was common.

One of the bases of engineering was projective geometry, introduced by Monge himself, which enabled three-dimensional objects to be represented in two dimensions with extreme precision. Before this formalization, technical drawing was based on perspective representations, which provided an overall idea of the machine as a whole and enabled the designer to understand what he had to build. After the introduction of projective geometry, the emphasis switched to the actual technical drawings, on which the various dimensions and the tolerances could be reported. The drawings allowed a worker to build the various parts of a machine, even a very complex one, without having a precise idea of the whole machine. It essentially separated the roles of the designer and the worker. Other intermediate roles dealt with structural calculations, the choice of materials, the economic aspects of production, the work cycles and so on.

This structure enabled increasingly complex machines to be built, and with mass production it was possible to build them at decreasing costs.

References

1. Francis Bacon, *The New Organon* 1, 129
2. M. Hulot, *L'art du turneur Mechanicien*, 1775.
3. *Ibid*.
4. Bergeron L.E., *Manuel du turneur*, Paris 1816.
5. Cipolla C.M., *Le macchine del tempo*, Il Mulino, Bologna, Italy 1981.
6. Farrington B., *Francis Bacon: Philosopher of Industrial Science*, Lawrence and Wishart, London 1951.
7. D'Alembert, *Traité de l'équilibre et du moment des fluides pour servir de suite un traité de dynamique*, 1774.

8

Technology, capitalism and imperialism: The rise of the West

8.1 EUROPE CONQUERING THE WORLD

Between the beginning of the sixteenth and the middle of the eighteenth century, Europe experienced an expansion that was nothing short of astonishing. In a short time, and without meeting particular opposition, the small states of the Old Continent gained control of the American and African continents, establishing a network of colonies. In the following century there came a showdown with the great powers of the Asian world and, at the end of the Opium Wars (1839–1860), even China was forced to accept the economic supremacy of the West. In less than four centuries, the relatively small area of Europe had taken over the whole planet.

In the twentieth century, the picture remained largely unchanged, except for the significant rise of the United States which saw a shift in the balance of power from Europe to North America. In this book, however, we choose to speak generically about 'the West', dealing with these two power bases as a single entity. In fact, even with all its distinctiveness from a technological and cultural point of view, American imperialism is a direct consequence of the European world, virtually free of any influences by the previous indigenous civilizations.

For many historians, the worldwide rise of Western imperialism was a sudden and surprising phenomenon. According to the majority of authors, in the period between 1500 and 1750 it would have been unthinkable to define a clear supremacy of Europe against the Asian powers[1]. The origin of the political, scientific and

---

[1] See, for example, Y. N. Harari, *From Animals*, cit.: "*During this era Europe did not enjoy any obvious technological, political, military or economic advantage over the Asian powers, yet the continent built up a unique potential, whose importance suddenly became obvious around 1850*".

technological primacy of the West can therefore be sought in this period, and the cause of the turning point is variously attributed to the sixteenth century Protestant Reformation or to the eighteenth century Age of Enlightenment.

Chronologically speaking, however, this reasoning is illogical. During these 250 years Europe had already launched itself into conquests of other continents, laying the foundations of modern Western imperialism. Those conquests were completed in a few years and with the use of only a small amount of military force; think, for example, of the blitz campaigns conducted by Hernán Cortés against the Mayan empire.

On closer inspection, the origin of European primacy is clearly an earlier phenomenon, yet, as previously mentioned, contemporary historiography refuses to conceive of a technological, material and intellectual development of Europe until 1492, intentionally placing itself in this illogical impasse. In fact, the imperialist thrust of Europe began during the upsurge in social, economic and technological transformations of the Middle Ages.

In addition to the aforementioned prejudices about the Dark Age, what leads many historians astray is that European civilization did not differ initially from its contemporaries because it had at its disposal a particular, clearly identified material technology. At this stage, the "European added value" was only based on its philosophical evolution. Beginning with Eriugena and the medieval technological revolutions, the Christian West had largely overcome its disdain for the earthly world and had developed a strong attitude of trust in the future and in the scientific discoveries of humankind.

The story of the fleet of the Chinese admiral Zheng He, who explored Indonesia and the entire Indian Ocean in the early decades of the fifteenth century, is emblematic [1]. He reached the Horn of Africa, the Persian Gulf, and the Arabian Peninsula, going as far as Mecca. Although the colonial potential of these territories was immense, in the years immediately following his last expedition, there was a change in the faction leading the government and the Mandarins suspended any further initiative. China went back to being a nation isolated from the rest of the world, and Zheng He's findings were quickly forgotten. The new imperial authorities, as well as those before the Zheng He era, considered the idea of new expeditions as a waste of resources and energy. Following the orders of the emperor, the fleet was destroyed.

This story mirrors that of the tenth-century steel boom mentioned earlier[2] which was also blocked by the conservative policies of the Mandarins, who cut short the industrial ambitions of the private companies by nationalizing the whole sector.

---

[2] See section 3.1 in Chapter 3.

In the fifteenth century, China clearly had the same material and technological resources as the European continent. What was different was the philosophical attitude of the ruling class, which saw only valueless cost rather than profitable investment in these expeditions. The exotic and risky expeditions of Zheng He had the sole objective of manifesting Chinese imperial power to the world, not that of obtaining a lasting economic advantage for the Chinese state. Even if, as some historians say, the expeditions were intended to find alternative trade routes to the Silk Road – at the time blocked by the invasion of the Tamerlane Mongols – the fact remains that the Chinese were unable to imagine a colonial network which could guarantee them sustained economic enrichment.

Compared to the millennial Chinese empire, Europe benefited from what Rodney Stark calls the *"blessings of disunity"* [2]. The competition among the many small states, which had emerged from the ashes of the Roman Empire and evolved over the course of a millennium, had made the ruling classes much more dynamic and open-minded. They did not hesitate to seek the most disparate and innovative solutions to ensure primacy over their uneasy neighbors.

Furthermore, unlike their Chinese contemporaries, the Europeans had clear objectives (see Fig. 8.1). As with the technological innovations, all the exploration missions were planned with the express desire to find new territories to exploit for the economic and material progress of the nations involved. In 1492, Cristopher Columbus landed in San Salvador for the first time, without even fully realizing that he had reached a new continent. Just thirty years later, Hernán Cortés had already completed the conquest of Mexico, despite the resistance of the Aztec empire.

Figure 8.1. World map by Cantino, made by a Portuguese navigator around 1502. All the unknown areas are left blank, also graphically demonstrating the European will to complete the exploration of the world. [Image from https://upload.wikimedia.org/wikipedia/commons/9/9c/Cantino_planisphere.]

The African conquests were completed even more rapidly. In less than fifty years, Spain had created a real empire which extended over three continents. Nevertheless, a new power was already rising, relying on a completely new form of economic organization.

The small region known as Dutch Flanders, freed from Spanish domination after a long season of conflicts, did not even have the necessary resources to undertake this type of colonial adventure. Nevertheless, they managed to rival the Spanish empire "*on which the sun never sets*"[3] thanks to the trust placed in them by the European financial system. The Netherlands did not finance its imperialist expansion with resources derived from military successes in Europe or from its own land holdings, but achieved success through the dynamism of its private companies and, paradoxically, thanks to the political and military weakness of its state.

Dutch merchants were considered to be particularly creditworthy because the local judicial system offered very little protection to insolvent debtors. Flanders was also considered financially safe because they did not have the military power to try to free themselves from their debts by resorting to force, as had often happened with France and Spain.

It was this imbalance between the small, private Dutch companies and the Spanish central state that made it preferable to invest in the former. Moreover, the empire of Madrid extended over a very significant portion of the European continent and was constantly compelled to squander its finances in local wars. These were very expensive campaigns with little or no economic gain, which drained most of the profits of the exploration missions in the new continents, to the despair of those who had invested their money in these enterprises.

The favor of German banking institutions allowed the Netherlands to supplant Spain progressively at a world-wide level and to found an empire that spanned all the known continents. The Dutch focused particularly on Africa, an inexhaustible source of raw materials, and the Far East, which until that time was still relatively unexploited.

From an economic point of view, it was actually a revolution. Even if they did not have the necessary resources to finance their expeditions, the Flanders commercial companies borrowed in advance what they needed from banks, subverting the millennia-old economic paradigm which stated that the procurement of resources must be done in advance of their use. This new idea can be considered an act of faith. A debt was subscribed in the present on the assumption that it would be repaid in the future, based on the outcome of an exploration mission into the unknown, as clearly seen on the incomplete maps of the sixteenth century.

---

[3] This expression was used by the Spanish emperor Charles V himself to describe the magnitude of his dominion. He ruled over the Asians isles of Saint Lazarus, that were later renamed 'Philippines' after his son, Philip II.

This implied an earnest confidence in progress which was also shared by the investors. In the Greco-Roman world, a commercial activity, however profitable, could never obtain credit as it was considered too risky and unworthy of financing.

The real obstacles of that earlier period, namely the disregard for the world of work and the lack of confidence in innovation, had long since been overcome. Now, the problem of risk had also been resolved by resorting to societies with a widespread share ownership. In order to avoid depending only on banks, these new subjects collected the capital they needed from a large number of investors, who acquired the status of members of the company in all respects. In this way, the possibility of loss was split across a large number of individuals. If the enterprise was successful, the profits remained substantial, because the colonial enterprises were extremely profitable and thus provided an enormous multiplier to the original investment, far superior to any other form of use of the same sum. Thus, a real stock market was started. If you wanted to withdraw your investment, there was the possibility of reselling your shares to another lender interested in joining the company.

Under these conditions, investing in this type of private activity became a reasonable bet. Since they were limited liability companies, in the event of a negative outcome the losses were limited to a maximum amount equal to the capital invested, with no possibility of additional expenses.

With the success of the first expeditions, the collective trust grew further and an increasing number of people, belonging to the upper classes, became interested in buying the shares of these exploration companies. Market prices rose and, at the time the new shares were issued, the companies received further benefits from a surcharge on the original value.

The largest joint-stock company of the time, the Dutch East India Company (VOC), was a real commercial superpower. With the profits from the stock market, it even managed to field a full-fledged mercenary army and conquer the Indonesian islands, which were ruled for over two hundred years by the Company as a direct possession.

While China was ruled by the authority of the central state at that time, European disunity was a fertile ground for the rise of private companies, which had a greater incentive for expansion than that of national states.

Even when the worldwide supremacy of the Netherlands was supplanted by the rise of the British Empire, the winning formula remained the same. The leaders of the colonial enterprises, which led the British to rule over one third of the planet, were private joint-stock companies, including the London Company, the Plymouth Company and the British East India Company. The latter even repeated the Indonesian campaign of the Dutch VOC, but on a much larger scale, obtaining power over the whole Indian subcontinent. Their possessions were nationalized by the crown only a century later, at the height of the Victorian age, but this change of rule was not a setback for the private colonial initiative. On the contrary, the imperial government was the authorizing power for these subjects who, as with today's lobbyists, held power and influence at court and influenced state policies.

8.2 SCIENCE AND EMPIRE

The four centuries of rapid political and commercial expansion of Europe completely transformed not only the other continents, but the Western world itself. Scientific and technological innovation was also deeply influenced by this crucial passage, which marks the beginning of the modern age.

The collective trust placed in these expeditions was not only based on the ability of the colonizers to obtain new wealth through colonial conquests, but above all on the possibility of discovering new resources and tools that could contribute to the advancement of European civilization.

Societies such as the Royal Society of London, which devoted themselves to funding the scientific component of the expeditions, were born alongside the exploration companies and they were not kept to the margins. This was an impressive effort that was carried out at the same scale as the colonial one. Over time, it became increasingly difficult to separate the two components, to the point that Harari speaks of a "*marriage of science and empire*" [3]. From this marriage came a substantial flow of novelties, discoveries and inventions which, to all intents and purposes, gave life to the First Industrial Revolution and to the contemporary world that we know.

In the wake of the enthusiasm generated by colonial expansion, there was even significant investment in general scientific research, even if it lacked immediate productive applications. For example, Captain James Cook's expeditions included experts of all kinds – astronomers, botanists, zoologists, anthropologists and physicians – and led to the discovery of cures for various diseases, including scurvy. From the British expeditions in the Pacific Ocean, Charles Darwin elaborated his thesis on the evolution of living species, while thanks to the subsequent expeditions in the Middle East, the archaeology and language of the ancient Mesopotamian civilizations were rediscovered after centuries of oblivion.

The study of plants experienced an extraordinary development. From an agricultural point of view, the expansion into other continents allowed the Europeans to import thousands of new products that revolutionized daily lifestyle, improving the diet, the life expectancy and individual well-being, and providing another of the foundations of the industrial age. In short, every scientific sector of the time, including technology, experienced significant improvement thanks to this constant flow of products, resources and data collected from previously unknown areas of the world.

What was surprising was the ability of the European colonizers to derive astonishing innovations using elements that had been available to the indigenous civilizations for centuries. When examining the distribution of raw materials, for example, it is surprising to note that Africa had a real treasure, capable of making the local populations extraordinarily rich and advanced. Significant deposits of gold, copper, cobalt, iron and many other metals essential for technological development were available, not to mention, from a more recent point of view, the abundant availability of uranium. Nonetheless, the indigenous civilizations could not benefit from this enormous potential and remained stuck for millennia in a technological stagnation that rejected *a priori* any kind of evolution. It was, as Karl Popper[4] would say, a *"closed society"*, blocked by the preservation of the *status quo*.

It is reasonable to say that, had there not been this imperialist push by post-medieval Europe, the resources in question would have remained largely unused, and humanity as a whole would not have known the extraordinary achievements of the contemporary age. In this sense, the widespread rhetoric of "western theft" from the rest of the world is a major oversimplification, if not an outright falsity.

8.3 PROGRESS GENERATES PROGRESS: A SYSTEM BASED ON TRUST

As noted with the Dutch colonial enterprises, from the seventeenth century onwards the entire European economic system was based on the collective trust in progress. Strange as it may seem, the true pillar of modern Western economics is this psychological factor.

While the ancient civilizations were characterized by slow growth, since the production volumes were proportional to the limited resources available, in the modern age it is possible to reverse this paradigm. Thanks to the intervention of external investors, it became possible to overturn the chronological order, also using in advance a portion of the outcomes that were expected to be obtained in their production, in spite of the fact they did not yet exist (Fig. 8.2).

---

[4] See K. Popper, *The Open Society and Its Enemies*, Routledge, London, 1945. At the very beginning of his work, Popper defined the closed society as a civilization that sees every mutation of the established order as a threat, and is particularly hostile toward scientists and inventors. On a political level, this society strives to preserve the *status quo*. Popper very effectively defines it as *"petrified"* (p. 45).

8.3 Progress generates progress: A system based on trust

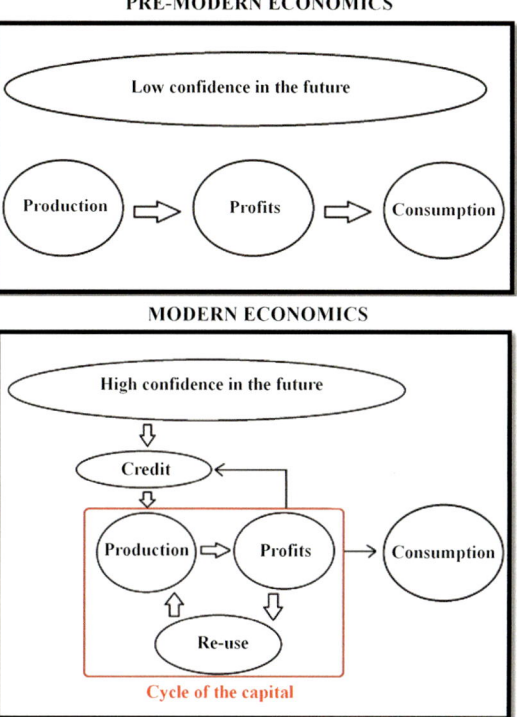

Figure 8.2. Pre-modern and modern economy: the role of capital.

The revolution did not arise from the invention of banks able to loan money, which already existed in the Athens of Pericles, but rather in overcoming the psychological barrier which at that time kept banks separated from the real economy: the lack of confidence in the future.

The true success of colonialism thus depended not only on the actual discovery of new lands and new resources, but also on the belief that these discovery journeys would bring enormous benefits to the West. Without this optimistic principle, which arose from both the first actual data and the philosophical influence of Christianity, no one would jeopardize their wealth to finance hundreds of expeditions. In the Eastern world, where for millennia there was a lack of confidence in the future and in innovation, such a leap would also have been hampered by their different view of the world.

While in eighteenth century China, as with the West at the time of the Roman empire, the economic trend depended only on objective elements such as raw materials and labor, in colonial Europe, external factors such as psychology and common opinion came into play.

With the emergence of public limited companies and the proliferation of potential investors, the economy increasingly became a democratic process, with all the advantages and defects that characterize it. For example, opinions could be inaccurate, or otherwise vitiated by prejudices. If everyone believed that a particular initiative was advantageous, then the banks and individual investors would finance the company proposing it, with the result that it would effectively have a greater chance of success. If instead they believed that it was not worthy of attention, or that it was too risky, then the company would not be able to acquire the needed resources. All this held true regardless of the accuracy of the opinions of those who invested, and therefore independently from the actual potential of the financed company.

The most obvious result can be found in the so-called 'asset bubbles'. As a result of an 'advertising campaign', *ante litteram*, from 1717 to 1719, the assets of the Mississippi Company enjoyed impressive success at the Paris stock exchange. However, none of the investors was aware of the fact that the region in question, today's Louisiana, was a swampy and unproductive area. When the truth emerged, the result was a drastic fall in the value of the assets and the annihilation of enormous sums of money, since there was no real evidence to support that kind of trust. Similar issues on the stock exchange, with devastating consequences on a global scale, most recently caused the financial crises of 1929 and 2007.

However, the most worrying scenario stems from the opposite hypothesis, which has represented a latent threat to Western civilization since the seventeenth century: the loss of confidence in the future. Although promising at the time of its initial devising, every technological invention of the last four centuries has always required financial risk from investors, if only at the time of its widespread dissemination. For every inventor whose work has entered history, there were lenders who believed so much in those inventions that they bet their wealth on them.

Why take the risk? Clearly, this type of conviction was not based only on the data available at the time. It was a real faith in progress, which derived from the medieval philosophical revolution. From Eriugena up to the present day, the drive of humanity to progress and grow has become a universally accepted fact, and any hindrance or economic loss represents only a "momentary setback". But what if the West loses this faith? As the economy is a trust-based system, there is a risk that this kind of pessimistic prophecy will turn out to be self-fulfilling, with devastating impacts on investment, consumption, and even scientific and technological innovation.

After all, the link between finance and science is now indivisible. If in Alexandrian Egypt the technologies represented only *mirabilia*, or "special effects" created to surprise and entertain the public, and in the early Middle Ages the *ars mechanica* still remained a sector in itself, albeit susceptible to practical applications on production, then starting from the seventeenth century, these tools become a cog in a much more complex machine which also included colonial expansion: the capitalist economy.

8.4 CAPITALIST THEORY AND THE ROLE OF TECHNOLOGY

Until the late seventeenth century, the economy was not a sector of particular interest. Not only is there no trace of previous literature on the subject – a fact which can easily be ascribed to the disdain of the aristocrats for economic reality – but it also turns out that no-one had actually ever dealt with it in any depth. In ancient Greece, *oikos-nomy* meant the household administration, the simple daily management of income and expenses. After all, in society at that time, the economic system did not go beyond balancing the income and expenses necessary to generate what was needed for subsistence for the majority of the population. For the wealthy, a surplus was produced merely to be accumulated, without any form of reuse.

"Why take the risk?", we wondered in the previous section. Given the lack of confidence in the future that was typical of the time, nobles preferred to accumulate income that exceeded what was needed for survival. The only alternative that was contemplated at that time was to spend them on other forms of consumption, even if they were unnecessary. From this behavioral dualism came the ethics of saving and prudence, which permeated all the classical literature: the wise man chose to accumulate, while the fool spent and wasted his possessions in the consumption of luxuries.

This was a very simple cycle that objectively does not require particular theoretical studies for its refinement. Even when some historians and philosophers occasionally dealt with topics that we would now call economic in the broad sense, such studies always remained on very general terms.

The first change was triggered by the experience of medieval monastic communities, where their vow of poverty prohibited any form of luxury or ostentation. Beyond a certain threshold, therefore, the accumulation of income made little sense, so they began to reinvest these resources systematically in the production process, thus generating a continuous, self-sustaining cycle.

A similar mentality also spread in the mercantile field, where a technical tool was developed to organize this cycle: double-entry accounting. At the same time, with the optimistic turn towards the future and the entry of credit into the economic system, the economy effectively became a complex system that allowed many possibilities and alternative solutions. Finally, in the seventeenth century, the euphoria for the success of colonial expeditions and the diffusion of private investment in public limited companies was the last piece of the gradual path of development for the Western economy.

As a result of these parallel transformations, the economy became a complex sector, deserving the interest of scholars. It is therefore not surprising that the study of this discipline was born in the West and in this particular historical period.

One of the first schools was represented by the so-called 'physiocrats', among whose leading exponents the French economist François Quesnay (1694–1774) must be mentioned. This first group of scholars wondered about the origin of the surplus – that is, the part of the income which exceeded the invested money and could be reused in the capitalist cycle – and came to the conclusion that this surplus could only come from agriculture. According to Quesnay, the technological-manufacturing sector would be nothing more than a zero-balance economic area, from which a surplus could not arise, and therefore did not deserve particular attention. This type of surplus would in fact be intrinsically linked to the multiplier effect that derived from the fertility of the soil, the only element that permitted a quantitative multiplication of the product[5].

Clearly, physiocrats were strongly conditioned by their belonging to pre-industrial France and had not yet elaborated a theory of value capable of quantifying economically the leap in quality that took place between the raw materials and the finished product. Essentially, for this economical school, technology was at best of secondary use.

In contrast, however, the first factories flourished in the British colonial empire, powered by steam and by new mechanical devices. These production centers did not multiply the raw materials but transformed them, producing a clear increase in their value. The Scotsman Adam Smith (1723–1790) understood the need to re-establish capitalist thinking. With his essay *The Wealth of Nations,* he laid the foundations of the study of modern economics [4].

According to Smith, the market itself generated the multiplier effect of wealth, which made it possible to increase the wealth of all producers, be they agricultural or industrial, with respect to the beginning of the production cycle. In fact, they transformed the products by increasing their value, and this increase was then transformed into money at the time of sale at market prices. At the same time, this individual enrichment was enough to keep the system in balance, with enormous benefits for everyone. Each individual further nourished the economic network with his own consumption, buying products and increasing collective well-being. Consumption and selfishness, in short, were the real engines of the Western growth.

In this scenario, technology was a key factor in maximizing the collective well-being. The use of machinery allowed the production process to be increasingly automated, and therefore workers became more specialized and their efficiency increased.

However, it was John Rae (1796–1872), another economist of Scots origin but much less well known than Smith, who fully understood the immense potential of the marriage between technology and capitalist production [5]. For Smith, economic growth depended only upon the accumulation of capital, the fruit of individual decisions. Technology was merely a useful aid in the optimization of these processes.

---

[5] For further information about this topic, see C. Napoleoni, *Smith Ricardo Marx: considerazioni sulla storia del pensiero economico,* Boringhieri, Turin, Italy 1970, pp. 31-48.

8.4 Capitalist theory and the role of technology

Rae, in opposition to many of his contemporaries, said that Smith had made a serious error with that assessment. According to him, economic growth depended almost exclusively on technological innovation, conceived by "men of genius" and applied with incremental improvements to the industrial reality. For Rae, technology became the cornerstone of economic progress, while all the other system variables were downgraded to the rank of secondary elements. This change of perspective also translated into a different perception of the role of public authority. Smith was a staunch supporter of *laissez-faire*, believing that the state should not interfere with market dynamics, whereas Rae was of the opinion that scientific research could not be delegated to private individuals only and that the public authority had to intervene heavily in this sector. This is the keystone of the capitalist system.

Initially, Rae's theories were not taken up, and were only rediscovered in 1848 when the famous British economist John Stuart Mill (1806–1873) published his *Principles of Political Economy*, in which he stated that *"in no other book[6] known to me is so much light shed, both on a theoretical and historical level, on the causes that determine the accumulation of capital"* **[6]**.

On the back of these studies, in the twentieth century, the Austrian Joseph Alois Schumpeter (1883–1950) further expanded upon this vital sector and for the first time highlighted a distinction between invention and innovation. Invention, of a purely scientific nature, consisted of the creation of new theoretical knowledge, while innovation meant the effective implementation of these inventions at a technological level, with an impact on production. The relationship between the two phases was evolutionary, with theoretical research producing applied technology, which in turn translated into economic growth.

Schumpeter also argued that the capitalist market could not be considered – as Smith had done – as a system in static equilibrium, since scientific and technological discoveries upset that scenario over time with a constant process of *"creative destruction"*. In his opinion, the economy was not based on price competition, but on the competition between innovations.

Was this merely a theory or could it be verified in reality? That question was answered in 1957 by the American Robert Merton Solow, who used a complex mathematical model to quantify the contribution of technology to the economic growth of the United States. The result of this study left no room for doubt. In the first half of the century, technological innovation had an impact on the increase in U.S. gross domestic product of 87.5 percent. The remaining 12.5 percent was the sum of the contributions of all other inputs, such as employment and the increase in raw materials.

---

[6] Mills is talking about J. Rae, *Statement of Some New Principles on the Subject of Political Economy, Exposing the Fallacies of the System of Free Trade, and of some other Doctrines maintained in the "Wealth of Nations"*, Hilliard, Gray and co., Boston 1834.

Ultimately, the modern Western economy is now a system completely dependent on scientific research and its ability to revolutionize everyday reality. We are not only surrounded by technology, which now permeates every aspect of our lives, but even the value of the contents of our wallet and that of our bank account depend almost exclusively on the progress of scientific evolution.

8.5 THE DARK SIDE OF THE CAPITALIST EMPIRE

In the collective imagination, imperialism and capitalism represent one of the darkest and most shameful pages in human history.

On the other hand, we have already seen that when speaking of the limited historical consideration given to medieval Europe – the birthplace of its extraordinary ascent – the contemporary West tends to take upon itself a millennium-old sense of guilt and to disavow its superiority with respect to other civilizations. This attitude is, as a rule, accompanied by celebrations of the alleged greatness of the Islamic Mediterranean and of the inventions of the civilizations of India, China and pre-Columbian America. Only the Greco-Roman civilization, unconsciously perceived as pre-Western, is not automatically included in this condemnation.

The reason for this self-inflicted condemnation can be attributed, according to current opinion, to the unfair suffering that the Western world inflicted on the rest of the planet thanks to the joint villainy of imperialism and capitalism.

At first glance, the term 'empire' evokes negative images. To fans of science fiction, it brings to the mind the tyrannical rule that dominates the universe of *Star Wars*, while in the political lexicon, the word 'imperialism' implies a despotic and cruel attitude. However, as observed by Yuval Noah Harari, *"empire has been the world's most common form of political organization for the last 2,500 years"* and provided a fertile ground for the cultural and scientific development of humankind [3].

In this case, we are not talking about a particular form of government. The Spanish empire was an absolute monarchy, the British one a parliamentary monarchy, while contemporary U.S. imperialism takes the form of a democratic republic, as was the case with the Athenian empire of the fifth century BC.

With the term 'empire', we mean only a political entity that brings together various heterogeneous populations under its rule and which has a natural propensity for expansion. As the case of the Mediterranean civilizations conquered by the Roman Empire demonstrates, this political organization tends to standardize the cultures of the subjugated peoples, allowing only those elements it considers particularly useful to remain. The objective of this process, which in the contemporary era we call 'globalization', is to homologate the various identities of the peoples subjected to the imperial rule, transforming them from foreign bodies to subjects integrated into a single organism.

8.5 The dark side of the capitalist empire

The process is slow and undoubtedly has a bloody component. In the case of the Romans, there were those who claimed that "*where they make the desert, they call it* peace"[7]. However, it is an equally incontrovertible historical fact that the conquered populations have also obtained enormous benefits from this linguistic and cultural homologation, as Roman globalization enabled the spread of technologies, innovations, knowledge and ideas developed thousands of kilometers away on all the shores of the Mediterranean Sea, for example.

In scientific terms, the benefit was immense. With the disappearance of political, linguistic and social frontiers, theories and tools could be circulated at a surprising speed, allowing a much larger number of individuals to introduce their own improvements.

The advantages deriving from the results of these innovations were even greater. The Roman Empire introduced roads, infrastructures, artisan techniques, food products and resources of every kind across the Mediterranean regions, causing a revolutionary leap in the daily lifestyle of areas such as the Iberian Peninsula and North Africa.

A similar argument, if one chooses to go beyond the politically-correct rhetoric, could also be applied to European colonial conquests. The British domination of India introduced an evolution that was hitherto unthinkable, with the diffusion of technologies, infrastructures and Western administrative models throughout India that even after many decades continue to represent the basis of the nation's contemporary success. Undoubtedly, the price of these benefits was a marked inequality between the periphery and the center, a distinctive feature of all empires that often results in strong social inequalities.

In short, the dark side of imperialism exists and cannot be denied. There are countless cases that could be brought to confirm the brutality of European colonizers, such as the wholesale killing of the primitive populations of Australia and New Zealand and the genocide perpetrated against the inhabitants of Tasmania. As in that case, the brutality of the European conquerors changed the balance in a region and approached something comparable to the "desert" left behind by the Roman "pacifiers".

However, we need to pay attention to the risk of oversimplifying a global phenomenon, which is actually very varied and complex. One of the most widespread arguments of anti-Western rhetoric is based on the loss of freedom for the colonized populations, which would otherwise have seen a peaceful life and self-government if they had not fallen under imperialist oppression. This argument fails to endure a deeper scrutiny.

---

[7] Tacitus attributes this famous quote to the Caledon general Calgacus, while describing his speech to his men just before a battle against the Roman legions. Whether this passage was actually a representation of the thoughts of the enemies of Rome, or an auto-criticism widespread among the intellectuals of the time, remains a controversial matter.

Of all the criticisms raised against imperialism, this is probably the most unfounded. Unlike what is commonly believed, Zulu Africa and the Aztec Mexico had long been organized according to the imperial model, and were ruled by particularly bloody and repressive regimes. In his military campaign, Hernán Cortés had the decisive support of many indigenous peoples who, exasperated by the cruelty of the Aztecs, immediately allied themselves with the new and unknown invaders to free themselves from this bloody oppression.

With all due respect to the Enlightenment myth of the "good savage", the Tenochtitclan empire used to sacrifice tens of thousands of men, women and children every year. On the occasion of the inauguration of the Templo Mayor, it is estimated that the number of ritual killings totaled more than 20,000 people, while during a normal day of celebration, the number could average around 2,000[8]. The purpose may have been seen as noble – to supply the gods with blood to maintain the cosmic order – but that does not alter the result.

Likewise, the history of Zulu Africa is a continuous succession of wars, prevarications, oppressions and sufferings which have very little to do with the idealized Eden assumed by contemporary intellectuals.

In almost every case, the original freedom which the independent population would have enjoyed before being subjected to imperial domination is nothing else but an identity myth, constructed retrospectively by peoples who, in reality, never had an actual independence. This is the case with the Palestinian Arabs, who for many centuries were subjected to the harsh regime of the Ottoman Empire before the English protectorate and then the birth of the State of Israel.

Over the past 2,500 years, the world's political order has almost always been punctuated by the uninterrupted succession of empires to other empires – at least using the term 'empire' as defined previously – and the rejection of this organizational system is essentially a utopian attitude, as there is no alternative model that simultaneously avoids the return of cultural barriers while also safeguarding contemporary scientific and technological progress.

Arguments in favor or against imperialism could continue *ad nauseam*: Europeans and Americans performed a countless number of actions on such a vast scale, resulting in enough material to document and support every sort of thesis.

The only two facts that are historically unquestionable are the following:

- the imperial organization is typical of humankind and represents a model that can certainly be perfected and humanized, but is very difficult to abolish completely;

---

[8] For further information on this topic, see for example I. Clendinnen, *Aztecs: an interpretation*, Cambridge University Press, Cambridge 1991.

8.5 The dark side of the capitalist empire

- without the proliferation and the affirmation of the imperial model, it would not have been possible to reach an amalgamation between cultures that provided the fertile ground for the development and diffusion of science and technology, factors that define humankind as we know it today.

The controversy is further complicated if we also consider the contribution of the economic component of Western imperialism, namely capitalism. As noted in the previous chapter, this term does not describe a particular alternative between the possible modes of production, but embodies the functioning of every complex economy. This is also a phenomenon that is easily criticized but difficult to abandon *a priori*, as it represents the natural evolution of the markets following the transition to the modern age. The role of state authority remains open to debate; one that, after registering the failure of the opposite poles (represented by radical neoliberalism and communism), continues in the contemporary political world. Regardless, it is certain that its principles, enunciated by Adam Smith and his successors, represent the natural logic governing the working of every modern state. The cycle of capital is an essential assumption of the contemporary world, without which the concept of technological investment itself would not exist.

Obviously, since it is a purely economic model, capitalism does not take into account the ethical sphere, and indeed its driving force – as Smith stated in *The Wealth of Nations* – comes precisely from the exaltation of individual interests.

This aspect of Western economy has contributed to throw a further shadow on European imperialism, because terrible acts have been committed in the name of profit against the populations subject to imperial rule. By reducing individuals to mere elements of a system, a capitalist organization tends to disregard the human dimension, and to perpetrate previously unthinkable actions on the basis of an icy monetary logic.

The most emblematic case can be found in the reintroduction of slavery in the colonial world. This practice, completely abolished in medieval times thanks to the decisive influence of Christianity, was reborn on a large scale at the time of the conquest of Africa, with the aim of optimizing the capitalist exploitation of American plantations. Unlike Roman slavery, it was an extremely efficient production model, a mechanism that was as inhuman as it was perfect.

Despite the clumsy attempts at disinformation carried out by contemporary anti-Catholic historiography, this regression of civilization occurred in spite of the strenuous opposition of the Roman Church, not with its blessing [7]. In 1435, for example, Pope Eugene IV openly condemned slavery in his Papal Bull *Sicut Dudum*, and ordered the Spaniards to restore *"to the previous liberty each and every one of those, of both sexes, who formerly lived in the islands called Canaries and were imprisoned since the time of their capture, whom they subjected to slavery"*, within a period of 15 days from receipt of the Bull. The Bull stated that the release had to be rapid and *"without any payment or acceptance of money"*, under penalty of immediate excommunication.

However, the order was punctually disregarded. The papacy had now lost its authority over the European powers, as confirmed by other Bulls on the same subject later issued by the popes Pius II and Sixtus IV, always with regard to the case of the Canary Islands. With reference to the European states, the threat of excommunication had now been reduced to a mere shadow.

In 1537, the great protagonist of the Counter-Reformation, Paul III, returned to the subject of slavery, which had reached a vast scale with the colonization of the American continent. In the Papal Bull *Veritas Ipsa*, the pope attributed the invention of slavery to Satan himself, claiming that *"the enemy of the human race, which always hinders all good men so that the [human] race could perish, has invented a way, never heard till now, with which it could prevent the salvific word of God from being preached to all peoples."*

In short, as Rodney Stark observed, *"the problem wasn't that the Church failed to condemn slavery; it was that few heard it and most did not listen"*. Even where attempts were made to put these directives into action it was impossible to prevail over the mechanisms of capitalism. In the Republic of Paraguay, from 1609, the Jesuits had created a group of communities intended to improve the living standards of the native Indians and to propose a social model that did not contemplate slavery. The attempt was immediately violently opposed by the Spanish and Portuguese slave owners who, in the mid-eighteenth century, took advantage of the weakening of the Jesuit order to conquer the communities one after another and regain control over the region.

This weakness was mostly due to the European-wide dissemination of laicism thanks to the Enlightenment, causing the Jesuit order to be seen only as an unfortunate articulation of the Church and one that interfered with the sovereignty of national states.

We can say, therefore, that it was in fact the celebrated secularization of the economy and the political liberation of the European powers from the authority of the Church that made the return of slavery possible. The practice was then definitively abolished only with the advent of the Industrial Revolution, when it was no longer required to maintain the capitalistic cycle.

8.6 THE UTILITARIAN DRIFT OF THE WEST

In the eighteenth century, during the so-called 'Age of Enlightenment', this deeply utilitarian scenario found fertile ground even in the philosophy of the time, which openly disowned the Christian inheritance.

Clearly, it was not an unambiguous phenomenon. For instance, Immanuel Kant, (1724–1804), while starting from an essentially agnostic metaphysical position, applied the Christian ethical message in a secular way by proposing the existence

8.6 The utilitarian drift of the West

of a categorical imperative; i.e. a universally valid moral law, summarized in the maxim "*act so as to treat humankind always also as an end and never simply as a means*[9]". Kant's works represent a milestone in secular ethics, a cultural phenomenon that in the last three centuries has undoubtedly provided a precious contribution both to keep the most ruthless directions of capitalism under control and to enrich the world of philosophy, science and technology.

At this time, a secularized humanism, repudiating the dogmas of religion and claiming the freedom of the individual to build his own moral rules starting from individual rationality, began to assert itself. Even though there was an evident discontinuity in the vision of the universe and of society, one can objectively observe how the system of fundamental ethical values did not undergo radical changes. The lay observer attributed this continuity to universal human reason, while the believer identified an underlying divine plan that manifested itself in the work of humans even without their knowledge. Different and not always compatible points of view, they nevertheless agreed in rejecting certain inhuman aspects of capitalism and found a precise point of convergence in the categorical imperative of Kantian morals.

At the same time, however, there were also those who saw in the human person only a means for achieving their selfish interest, providing further support to the most ruthless aspects of capitalism.

This was true, for example, of the famous Marquis Donatien-Alphonse-François de Sade, (1740–1814), a diehard atheist and advocate of individual freedom in all its forms who, in his *Juliette*, proposed a complete disavowal of the traditional Christian ethics. Refuting Kant's categorical imperative, the Marquis de Sade provocatively claimed the right to use people to attain the only true ultimate end of nature, namely pleasure: "*I must be able to enjoy whoever I want, regardless of whether the person can suffer.*" As Roberto Giovanni Timossi observed, de Sade's thinking was actually the natural outcome of a mindset that repudiated the existence of a deity: "*If there is no God, there are no spiritual substances, there is only matter and therefore bodily pleasure is the only true purpose of human existence. For de Sade, you can't be atheists and can't be immoral*" **[8]**.

---

[9] *Foundations of the metaphysics of morals*. Kant maintained that humanity should be regarded "also" as an end. He quietly implied that it could be considered as a means as well, as long as it was not exclusively regarded in this fashion. Kant, as a philosopher of the so-called 'Age of the Enlightenment', did not fully agree with Christian morality, which instead regarded mankind always as an end and never as a means. According to the Book of Genesis and the Gospel as well, every man is the living image of God, and therefore is sacred.

Unlike what was suggested by his many biographical misadventures, de Sade's thinking was successful and deeply influenced modern society. At a distance of over two centuries, even the contemporary philosopher Joel Marks argues that *"in the world there are no literal sins because literally there is no God and hence the whole religious superstructure that would include such categories as sin and evil. Nothing is literally right or wrong, because there is no morality"* [9].

The atheist Marks believes that the only truly consistent position for those who do not recognize the existence of a deity must be to repudiate any morality. This position is obviously quite radical and not widespread, but it sheds new light on the strong seventeenth-century utilitarian drift that underlies many contemporary injustices.

At an economic level, however, with the disappearance of the sacredness of the human-creature, there is nothing left but the private individual, often abandoned at the mercy of the selfish interest of those who operate the market. Such interest, while theoretically ensuring the growth of the collective well-being of which Adam Smith spoke, does not take into account the protection of individuals and tends constantly to increase social inequalities.

By exalting the individual's full freedom of action, the freedom to act to the detriment of others is therefore also gradually accepted, thus paradoxically passing from an excess of individual freedom to authoritarianism. With Friedrich Nieztche (1844–1900), the idea of the affirmation of the "lust for power" of individuals who were superior by nature to the "flock" that surrounded them was also introduced to philosophy, paving the way for potentially more dangerous outcomes.

Politically, such a trend was also the basis of the great totalitarianism of the twentieth century, which apparently exalted the interest of the community to the detriment of individual protection. Nazism, Communism and Fascism are nothing more than a broader transposition of this extremist individualism, which celebrates the absolute freedom of a single individual – or people – to take decisions at the expense of everyone else.

If the solidarity ideal of the Christian message is abandoned, without substituting it with similarly strong and well-founded secular ethics, one can even make a case for legitimizing the physical elimination of the opponent, as long as it falls within the collective interest of the race or social class or, at the economic level, of the market.

Even the French Revolution itself experienced a period of systematic killing of the "enemies of the nation", which was implemented by the Jacobin Club that most closely embodied the extreme secularism of that era. This obviously refers to the well-known period called the Reign of Terror of Robespierre, but also to the violent persecution of the ecclesiastics and to the systematic destruction of the sacred art of Avignon. In reality, one of the first casualties of the French guillotine was the categorical imperative of Kant and its straightforward defense of

8.6 The utilitarian drift of the West

humanity at all costs. The Reign of Terror had very little to do with the balanced and sober "state of right" imagined by the German philosopher, who initially strongly supported the revolution[10].

It is evident that the use of violence and abuse against opponents is a phenomenon as old as humankind, and certainly cannot be attributed to any particular political current, or religious or philosophical thought. However, there is no doubt that the overlap between Adam Smith's individualism and the disappearance of the millennia-old Judeo-Christian morality contributed greatly to loosening the ethical boundaries of the West, in both economics and in politics. The result was the advent of a season of exasperated utilitarianism, exposing the inhumane face of imperialism and capitalism. This is a face that, unfortunately, continues to manifest itself today to the detriment of the poorest and most vulnerable populations on the planet.

Not surprisingly, technology has also played a central role in the violent affirmation of this 'dark side' of the West, especially with the transformation of scientific achievements into weapons of mass destruction in order to maintain its rule over subjugated populations. But are we really sure that the blame must fall on these technological tools?

References

1. Finney B., *The Prince and the Eunuch*, in *Interstellar Migration and the Human experience*, University of California Press, Berkeley, 1985.
2. Stark R., *How the West Won. The Neglected Story of the Triumph of Modernity*, Intercollegiate Studies Institute, Wilmington, Delaware 2014, p. 112.
3. Harari Y. N., *Sapiens. From Animals into Gods: A Brief History of Humankind*, Harper, New York 2015.
4. Napoleoni C., *Smith Ricardo Marx: considerazioni sulla storia del pensiero economico*, Boringhieri, Turin, Italy 1970, pp. 49-93.
5. See Coccia M., *Le origini dell'economia dell'innovazione: il contributo di Rae*, Ceris-Cnr: Istituto di Ricerca sull'Impresa e lo Sviluppo, working paper 1/2004, pp.1–19.
6. *Ibid*, p.15.
7. Stark R., *Bearing False Witness: Debunking Centuries of Anti-Catholic History*, Templeton Press, West Conshohocken 2016.
8. Timossi R.G., *Nel segno del nulla*, Lindau, Turin, Italy 2016, p. 367.
9. Marks J., *An Amoral Manifesto* in *Philosophy Now*, online, 2010.

---

[10] The philosopher V. Mathieu summarize the matter in *La rivoluzione politica e la libertà di Kant,* online, p. 49, as follows: "*the change Kant hoped from the revolution was the establishment of a republican constitution, as theorized by Montesquieu: he didn't think about the abolition of the monarchy (Kant, being a loyal subject of the King of Prussia, would never dream of something like this), but a republic ruled according to the needs of the people, as we would say, instead of the needs of a single person, a dynasty or any social class*". The utter silence about the so-called Reign of Terror from Kant, who had been a firm supporter of the revolution in 1789, has been largely interpreted as a sign of embarrassment toward a political phenomenon that had dramatically changed since its beginning.

9

The dark side of technology

9.1   WEAPONS AND DUAL TECHNOLOGIES

Weapons appeared very early in human history. Indeed, we can say that technology began with weapons, at least those for hunting. It has been said that, from the beginning, humans felt the need to complete their diet with proteins from animals, but their physical structure was never designed to be a natural predator. Therefore, they had to resort to tools to make up for this serious handicap.

Obviously, we do not know when a human first turned these hunting weapons against another human, but it certainly happened in the most remote antiquity. In the Bible, this event is even traced back to Cain, one of the sons of Adam, while paleontologists have found human remains that bear traces of wounds inflicted by weapons. Even if it is archaeologically impossible to distinguish between a deliberate homicide and a hunting accident, what is certain is that all the technologies of the stone age can be defined using a modern term: dual use. Any stone instrument could also be used as a weapon.

Moreover, there were very few in the past who had ethical reservations on this matter. When he listed the technological achievements of man, St. Augustine stated quite matter-of-factly that he *"invented all kinds of poisons, weapons, tools against men themselves"* without apparent moral discomfort. However, there can also be little doubt that the availability of weapons can facilitate civil coexistence, preventing stronger or more cunning people from always having the best and disposing of others as they wish.

As has been said several times, the progress of technology has led to the creation of a growing disproportion between actions and their effects. As far as weapons are concerned, the first such discontinuity occurred with the introduction of fire. The destruction that can be caused by setting a fire is absolutely disproportionate to the extent of the required action. But perhaps the most significant early discontinuity came with the invention of the crossbow. The aforementioned ban against the use of this weapon was a symptom of the unease it had caused. The crossbow gave rise to the possibility that a second-rate soldier, probably with minimal training, could kill a warrior, perhaps powerful and of high social class (given that the dart of a crossbow could pierce armor of the type only the powerful could afford). It meant that nobody, not even the nobles or kings, could feel relatively safe on the battlefield any longer.

9.2 MASS DESTRUCTION IN ANCIENT TIMES

It is often noted that, in the twentieth century, war itself could be considered more barbaric. Up to World War I, the greatest number of victims of any war had always been registered among the soldiers, while by the time of World War II, civilian casualties outnumbered those among the armed forces of both sides. The reason behind this can be traced back to two of the technologies which we described in the previous chapters as great achievements of humankind: aviation and nuclear energy (even though only two nuclear weapons were actually used). These two technologies brought about what are normally considered to be 'weapons of mass destruction'.

Actually, the problem here is not so much the extent of the destruction caused as the ease with which the technology enabled it; in other words, the disproportion between the entity of the causes and that of their effects.

By reading the descriptions in the *Odyssey* and in the *Aeneid* that Homer and Virgil provided of the fall of Troy, it is easy to realize how the looting of a vanquished city was an extremely dramatic and bloody event in ancient times, certainly no less so than a modern bombing. The only difference was that the victorious warriors chased and killed the inhabitants one by one, but did not dispatch all of them since some, particularly women and those who were worth more alive than dead, were used in other ways. During World War II, the operation of a bomb release button was both more clinical and more indiscriminate.

Before the introduction of agriculture, when slaves did not yet have an economic value, the only reason it was worth taking prisoners was to keep them alive, to eat them later. The description that Jared Diamond gives of the treatment that the Maori reserved, up till recent times, for the populations they conquered is emblematic of this stage of human civilization [1].

But there is more. The rights of the winners were not limited to killing or taking the vanquished population as slaves. Looting went far beyond taking possession of their properties and their lives, and often included the most total destruction of their territory. The victors burned the houses and the crops, cut down the fruit trees, uprooted the vineyards, killed the cattle and destroyed the dams and the other works that were necessary to an agricultural society. All this can be summed up in the sentence *"reduce fertile land to pasture for sheep and goats"*. We have reliable evidence of this practice in both Greek and Roman times.

In many cases though, it was the retreating army that destroyed the crops and dwellings of its own population, to prevent the enemy from finding logistical support in the occupied territory. The so-called 'scorched Earth' strategy is very old.

The right of devastation was universal, perhaps with few exceptions, but no one knows how reliable. The Greeks were amazed that, at the end of a war in India, the victors did not destroy the cultivated fields of the vanquished population. However, the knowledge that the Greeks possessed about Indian customs was quite limited.

Even those few who opposed this practice did not do so for ethical reasons, but simply because they did not consider it strategically convenient. A passage by Polybius (206 BC – 126 BC), who in addition to being a famous politician and historian was also the inventor of a practical optical telegraph, is symptomatic of this, as reported by Karl-Wilhelm Weeber: *"I do not like to approve the behavior of those who, out of anger towards their own people, not only end up stealing from the enemy the harvest of the year, but also destroying the trees and farms with all their equipment, without leaving any space for repentance: Those who act in this way commit, in my opinion, a serious error, to the extent that, with the devastation of the fields and the destruction of the foundations of existence, they terrorize the enemies not only for the present, but also for the future (...), they provoke them to the greatest extent, and if those had hitherto been limited to doing evil, now they will act out of irreconcilable hatred."* [2].

Such destruction of the fields and harvests did not require purpose-built technology, although Plutarch informs us that the Spartan king, Cleomenes III, with what the historian himself called *"uncommon efficiency and power of intellect"*, had his soldiers equipped with appropriate wooden clubs in the shape of large swords to break up the wheat from the fields around the city of Argos. However, fire would normally suffice for such destruction.

Fire could also be used with even more destructive effectiveness, if a firestorm could be unleashed. This is a phenomenon that can also occur naturally in forest fires. Above the fire, the hot air rises and draws air from the surrounding areas, creating very strong winds that fuel the fire and cause very high temperature increases of up to 1,500°C.

In his campaign in Gaul, Julius Caesar decided to set fire to a camp where an entire tribe of Gauls had settled. His legionnaires placed bundles of dry branches around the camp and then set them alight. It is not known whether someone had evaluated the potential effect or whether it was just a coincidence, but the conditions were right to trigger a real firestorm, and there were no survivors. In one night, between 100,000 and 150,000 Gauls died, warriors and civilians included. It is certainly possible that Caesar had not thought that his strategy would work so well, but it is to be believed that if he had known this, he would not have hesitated to cause such a massacre anyway. On other occasions, he is known to have destroyed entire tribes without sparing anyone, as in 58 BC near Grenoble, when he killed 130,000 Helvetii warriors plus all the women and children following them (whom he did not even bother to count), or in 55 BC when he massacred a group of 400,000 Germans on the banks of the Rhine[1].

A firestorm can easily be triggered by an intensive bombardment, particularly if incendiary bombs are used to start the fire. A well-known case is that of the bombing of Dresden in 1945. The number of victims is disputed, but an investigation that began in 2004 and ended in 2010 established that there were between 22,700 and 25,000 casualties **[3]**. Much higher estimates have also been circulated, of between 200,000 (the figure provided by the Nazi propaganda ministry) and 250,000 (the figure provided by the German government after the war). Such high numbers seem unfeasible, given that the total number of victims of bombings in the entire war and on the whole of German territory is only estimated at 370,000. This would mean that throughout the war the figure for victims of bombings in the whole of Germany was actually less than the number of victims of just one of the many massacres perpetrated by Caesar in the Gaul war. It has also been debated at length whether the bombing of Dresden should be considered as a war crime.

9.3 ANCIENT BACTERIOLOGICAL WARS

Chemical and bacteriological weapons are considered particularly heinous weapons of mass destruction, and they are often mentioned among the worst products of modern technology. While there are few historical precedents for chemical weapons (even if the practice of poisoning the enemy's wells has been repeatedly used in history), as far as bacteriological warfare is concerned the precedents are there, and they are particularly serious.

The 'weapon' was the same in all cases. In 1346, the Tartars besieged the city of Caffa, a Genoese commercial base in Crimea at the western end of the Silk Road. A few years earlier, around 1330, a plague epidemic had broken out in East Asia, which had spread to much of the continent and had also infected the Tartar tribes of southern Russia. Shortly after the outbreak of the plague among the Tartar army, their Khan Janibeg ordered the use of catapults to launch the corpses of the plague victims into the besieged city. As observed by the notary De' Mussis, a Genoese notable present in the city, "... *soon the stench of the decomposing bodies plagued the air ... it poisoned the springs and the stench was so oppressive that it was not long before barely one man in a thousand was capable of escaping what*

---

[1] These figures, which are very high indeed, are taken from E. Durschmied, *From Armageddon to the fall of Rome*, Hodder & Stoughton, London, 2002.

remained of the Tartar army" **[4]**. De' Mussis simply spoke of the stench of decomposition, but we now know that what spread through the besieged city was *Yersinia Pestis*, the bacterium that caused the disease.

There followed scenes similar to what we later saw in Saigon in 1975, with people storming the ships in the harbor in a desperate attempt to escape the epidemic. Unfortunately, the most effective carrier of the contagion, the rats, also got aboard the ships, and the epidemic spread across Europe starting at the ports in which those fugitive-laden ships docked.

The Black Death spread throughout the known world and it was calculated that it caused about 200 million deaths, a catastrophe that, as far as the number of victims is concerned, is second only to World War II. If we take into account that the world's population was much smaller in 1300 than in 1940, in relative terms it was the worst catastrophe humanity has ever seen.

We certainly do not mean to infer that this act of bacteriological warfare by the Tartars caused the Black Death epidemic in Europe. In all probability, the epidemic would have arrived on the European continent anyway and the health and social conditions of the time would have allowed it to spread, although perhaps in a slower and less destructive way.

What happened during the siege of Caffa was probably not the first case of this kind and the practice of infecting besieged cities using the corpses of plague victims was repeated several times in ancient history. It cannot be said that any scientific technology was necessary to implement such war techniques. It was enough merely to know that plague is a disease that can be transmitted by contact with infected corpses and this at least was understood, even if nobody knew how the contagion occurred or whether it was believed that infectious diseases were not spread by contagion but by more mysterious or supernatural sources. Of course, the other major requirement was the artillery capable of throwing objects of some tens of pounds (or less, given that only objects that had been in contact with the plague victims needed to be used). The point we are highlighting here is that weapons of mass destruction were in use well before modern weapons, which we believe to be a 'poisoned fruit' of modern technology, were developed.

9.4 NUCLEAR WEAPONS

Turning to nuclear energy, we have already noted that scientific research in this field was hastened and strengthened at the beginning of World War II by the drive to introduce explosives much more powerful than any existing at the time. In fact, what drove many physicists, particularly the Americans and British, to try to convince their governments to build the nuclear bomb was more than the desire to use this potentially very powerful weapon. It was the fear that Nazi Germany had undertaken similar research and could be the first to be equipped with such devices.

Nobody had the slightest idea precisely how powerful nuclear weapons could be and, above all, nobody was truly aware of the effects of the radiation released by the nuclear explosions. This underestimation of the danger, not unlike what had happened half a century before with X-rays, caused some serious accidents during nuclear tests in the 1950s.

In underestimating the effects of radiation, in the minds of the same researchers working in that field a nuclear bomb was simply a bomb that caused a much more powerful explosion than a conventional one.

In August 1945, two nuclear weapons were actually dropped on the Japanese cities of Hiroshima and Nagasaki, the only nuclear weapons that have ever been used. The first caused between 100,000 and 200,000 casualties and the second approximately 80,000, including those who died in the subsequent days and months due to radiation exposure, although it is impossible to supply precise estimates. These figures are very similar to the casualties caused by large-scale conventional bombings, such as those of Dresden, Hamburg or Tokyo. The bloodiest battle of the Pacific, at Okinawa, caused 120,000 deaths among the Japanese and 18,000 among the Americans[2].

The strategic and ethical evaluation of the use of the nuclear weapons is controversial. There is no doubt that the explosion of the two nuclear bombs forced the Japanese to surrender and ended World War II, and it is equally certain that if the bombing had not been performed, the Japanese government would have decided to continue the war for many months. It was also calculated that the willingness of Japan to continue the war would cost the Japanese about 200,000 civilian victims a month. To the direct victims of the continuation of the war, indirect victims must also be added. The submarine blockade, the mining of the Japanese coasts and the paralysis of transportation due to the bombing of the railways would have caused a famine that would have led to death by malnutrition for a large number of Japanese. Here too, the estimates are difficult and certainly not reliable, but figures have been suggested ranging from hundreds of thousands to ten million, the latter figure provided by the historian Daikichi Irokawa **[5]**.

Some Japanese historians have claimed that the civilian leaders, who were in favor of surrender, saw their salvation in the atomic bombings. Koichi Kido, one of the closest advisers to Emperor Hirohito, declared: "*We at the peace party were helped by the atomic bomb in our attempt to end the war*". Hisatsune Sakomizu, the chief cabinet secretary in 1945, called the bombings "*a golden opportunity given by heaven to Japan to end the war*".

American estimates of the losses their armed forces would have suffered in invading Japan range from 100,000 to half a million. Even this is probably a miscalculation, given that the residual Japanese forces had been underestimated by

---
[2] The figures and the declarations reported here are mostly taken from https://it.wikipedia.org/wiki/Bombardamenti_atomici_di_Hiroshima_e_Nagasaki.

the Americans by a factor of about three. In addition, we must consider a probably significant percentage of the 600,000–800,000 prisoners in the Japanese concentration camps in Asia, who would not have survived a prolonged war, as well as the many Asian civilians, especially Chinese, who were indiscriminately killed by the Japanese, and the prisoners of war who would certainly have been killed as a result of an executive order of the Japanese war ministry to execute all prisoners should Japan be invaded.

Although it is ethically dubious to make calculations of this kind, those who support the moral legitimacy of the bombings of Hiroshima and Nagasaki believe that those bombings saved a substantial number of lives.

John A. Siemes, a professor of modern philosophy at Tokyo Catholic University and witness of the atomic attack on Hiroshima, wrote: "*We discussed the ethics of the use of the bomb among us. Some considered it in the same category as poison gases and were against the use on the civilian population; others were of the opinion that in the total war, as was carried out by Japan, there was no difference between civilians and soldiers, and that the bomb itself was an effective force that tended to put an end to the bloodshed, warning Japan to surrender and thus avoiding total destruction. It seems logical to me that those who supported total war since the beginning cannot complain about the war against civilians.*"

Actually, at the time there were few voices of dissent against the American action. In Italy, for instance, *L'Unità*, the official organ of the Italian Communist Party at that time, published an article a few days after the bombing entitled *Serving civilization*. The article said, among other things, "*The news that the U.S. Air Force used the atomic bomb has been accepted in certain environments with a sense of panic and with words of reprobation. This seems to us a strange psychological complex, a formal obedience to an abstract humanitarianism*".

In contrast to these favorable considerations, there was also much criticism. Even before the actual use of the atomic weapon, some scientists who had supported the Manhattan project, in particular Albert Einstein and Leo Szilard, believed that while the development of the nuclear weapon was necessary initially when it was likely that Nazi Germany was engaged in such a project, once Germany had fallen this urgency was less valid. Above all, they considered that the United States should not be the first to put atomic weapons to use. The two scientists wrote a letter to President Roosevelt about the issue, and Szilard went on to say: "*If the Germans had thrown atomic bombs on the cities in our place, we would have called the release of atomic bombs on the cities a war crime and we would have condemned to death the Germans guilty of this crime in Nuremberg and we would have hanged them.*"

Even Edward Teller, who was later the main supporter for the construction of the thermonuclear bomb, argued that the nuclear bomb should have been dropped on an uninhabited area of Japan to convince the Japanese government to surrender without causing a large number of casualties.

Some jurists have argued that all such attacks on civilians were contrary to the various Hague conventions, ratified by the United States, Article 25 of which states: "*It is forbidden to attack or bomb, by any means, cities, villages, houses or buildings that are not defended*". This accusation was accompanied by the statement that a bombing conducted to convince a state to surrender is state terrorism. However, based on these criteria, any bombardment at all would qualify as state terrorism and would be a war crime, regardless of whether conventional or nuclear weapons were used, and also regardless of the extent of its effects.

Many military leaders, including General Douglas MacArthur, Admiral William Daniel Leahy, General Carl Spaatz, Brigadier General Bruce Clarke, Admiral Ernest King and Admiral Chester Nimitz, later declared that there was no military justification for the bombing, as President Eisenhower also wrote in his memoirs.

Other arguments against the use of the nuclear weapon are based on the belief that, in reality, the Japanese were prolonging the war only to obtain more favorable conditions for peace, and that they would have surrendered anyway in a short time. However, these are just conjectures, and nobody can say for certain what would have happened. Another factor to be considered was that the Japanese government, whose military insisted on resisting to the end while other exponents were in favor of surrender, remained deeply divided. A further accusation is that the bombing would actually have been conceived more as a warning to the Soviet Union, in view of the forthcoming and predictable conflict that might have evolved in the Cold War, than as a pressure on Japan to obtain its surrender.

In a report of the National Council of Churches from 1946, entitled *Atomic War and Christian Faith*, it states: "*As American Christians, we deeply regret the irresponsible use already made of the atomic bomb. We have agreed that, whatever the judgment we may give about war in principle, the surprise bombings of Hiroshima and Nagasaki are morally indefensible.*"

A further criticism which can be levelled against the use of nuclear weapons in the last phase of World War II is that their actual use has led a section of public opinion to demonize not only nuclear weapons and their use, but nuclear energy in a broad sense, and in many cases even modern technology in general, giving rise to the anti-nuclear and anti-technology movements.

Up to this point we have talked about the actual use of nuclear weapons. However, their purpose, at least in the minds of those scientists who worked to bring them to fruition from 1939, and who opposed their use in 1945, was not to build them to use in place of conventional explosives, but to hold them as a deterrent that would make the prospect of war so scary that it in fact became impossible.

In this sense at least, nuclear weapons have fulfilled their task, perhaps even beyond the expectations of those who had promoted them. In the last phases of World War II, when the Soviets advanced in Germany and occupied a considerable part of Europe, all the military and political leaders assumed that a third world war would follow more or less immediately, fought among the winning powers, and in particular between the United States and the Soviet Union, the only powers still able to fight.

The fact that the United States possessed nuclear weapons made this war virtually impossible. Again, this comes back to the accusation that the United States' bombing of Hiroshima and Nagasaki also served to demonstrate the appalling power of these weapons to the Soviet Union.

For a few years, the United States was indeed the only nation to possess nuclear weapons, but this imbalance ceased in 1949 when the Soviet Union carried out their first nuclear test. From that moment on, what became known as the balance of terror was established, and war between the two superpowers became practically impossible. The creation of thermonuclear weapons, even more frighteningly powerful, made this stalemate permanent and led to a period of peace among the superpowers that still lasts at present, and which has almost no precedent in history.

Over the years, nations such as Britain, France, China, India, Pakistan, Israel, North Korea and South Africa have also armed themselves with nuclear weapons (although South Africa later renounced them), while other nations are suspected of possessing or of working to build nuclear weapons. However, various treaties have been signed: for non-proliferation to prevent other countries from developing these weapons; for banning nuclear tests in the atmosphere to prevent damage from radiation in the surrounding environment; and for limiting nuclear weapons in which the parties subscribed to limit their number.

9.5 DUAL TECHNOLOGIES

Dual technologies are technologies that can be used in both the civilian and military spheres. Essentially, almost all the technologies qualify, beginning with those developed in the stone age. From this point of view, fire is one of the oldest dual technologies, and was certainly the most important in antiquity.

However, here we mainly deal with those modern technologies that have been developed for, or at least supported by, military goals and have then become important in the civilian field. Among them, we must not only list nuclear energy for peaceful use (directly derived from the military use of nuclear energy), aviation (which was very strongly backed by the military), and space launchers (the direct descendants of ballistic missiles), but also computers, telecommunication systems, radar, the Internet and countless other technologies that today we consider to be purely civilian. Their military use has long since lost its importance, even though they were developed for that purpose initially. Even some scientific developments, which were later turned towards pure science, were devised for military use. One important example of this is the guidance and navigation algorithms used in space navigation, which were originally developed for intercontinental ballistic missiles. Another is cryptography, which today is so important for cellphone technology but was born and developed for military needs.

An example of the use of military funding to develop a civilian technology, dating back to the 1950s, is that of commercial jet aircraft. This has radically changed air transportation, particularly long-range transportation, making it cheaper, safer and more comfortable and thus enabling a significant portion of the world's population to make intercontinental trips, which were previously the reserve of the particularly wealthy elite. The first civilian jet aircraft to be commercially successful, the Boeing 707 (the earlier De Haviland DH 106 Comet initially had serious safety problems, and, when these were solved, failed to become a commercial success) was developed only thanks to the experience that Boeing had gained from building large military aircraft such as the B-47 Stratojet bomber, the military transport C-97 Stratofreighter and the military KC-135 Stratotanker [6]. Therefore, it was military funding that enabled this leap forward in civil aviation. A similar approach was attempted for the development of commercial supersonic airplanes, with the military funding of the supersonic bomber B70 Valkyrie. This was intended to be used to develop the prototype SST (Super Sonic Transport). In this case, however, the approach did not work, as the development of intercontinental ballistic missiles made the idea of strategic bombers obsolete, and project B70 was cancelled. This also caused the cancellation of projects intended to create civilian supersonic aircraft. The only project to survive this was the Anglo-French Concorde, which was born under the pressure of national prestige of the two nations involved and did operate for several years, though it never really became operationally convenient.

Another example is the current trend to create autonomous motor vehicles, which initially saw quick progress thanks to the considerable funding of DARPA, the U.S. Defense Advanced Research Projects Agency.

Similar trends can be seen when examining past history. Was Galilean mechanics itself not born around the problem of the motion of projectiles? And were the machines of the Renaissance "theatres of machines" not for the most part designed for military use? Looking to the more remote past, has the research of Archimedes on optics not become famous for the use – though we do not know how true this might be – of burning mirrors in the war against Rome?

Returning to the present, the significant development of science in the United States has regularly been boosted by the commitment of the army, the navy and the air force in promoting basic research without worrying whether any specific field could have a direct military use, in the belief that knowledge in itself has a fundamental impact on national security. In the United States, the Army Research Office is one of the major funding organizations for basic research. Around the world, such research receives very limited public funding.

182 The dark side of technology

This mixture of civilian and military use of ancient and modern technologies, and of many applied sciences, suggests that it is improper to speak of a 'dark side' of technology. From this point of view, technology is 'neutral' and simply provides possibilities. As always, it is humans that take the decision about what use to make of the tools that technology puts at their disposal. It is up to them to make responsibly ethical choices.

References

1. Diamond J., *Guns, Germs and Steel: The Fates of Human Societies*, Vintage Publishing, New York 1997.
2. Weeber K.W., *Smog über Attika*, Artemis Verlag Zürich and München 1990.
3. *Dresda, i morti furono 25 mila,* in archiviostorico.corriere.it, online, 18 marzo 2010.
4. Kelly J., *The Great Mortality: An Intimate History of the Black Death, the Most Devastating Plague of All Time*, Harper Collins, New York 2005.
5. Daikichi Irokawa, The *Age of Hirohito, In Search of Modern Japan,* Free Press, 1995.
6. Heppenheimer T.A., *A Brief History of Flight*, Wiley, New York 2001.

10

Industrial revolutions

10.1 THE AGRICULTURAL REVOLUTION

At the beginning of the eighteenth century, Britain introduced agricultural innovations designed to increase labor productivity significantly. They reduced the number of workers engaged in agriculture and freed up the manpower which subsequently enabled the Industrial Revolution to take place. Actually, agriculture had already seen productivity increases in the past, but these were mostly volumetric, either due to increasing the amount of land used for cultivation, or by increases in population and therefore available manpower. The actual *rate* of production essentially remained constant. In 1700, in contrast, the significant increases in productivity were accompanied by the decrease in the agricultural manpower, with the consequence that total production did not increase more than the population and there was a stagnation in the availability of food per capita.

What was later termed 'the agricultural revolution' began in 1701, when Jethro Tull invented and perfected a seeder that allowed the seeds to be distributed uniformly and at the desired depth. This invention lasted for more than a century, although Tull's seeder was rather expensive, somewhat unreliable and did not have a major impact on agricultural production. More important, in 1730, was the introduction by Joseph Foljambe of the 'Rotherham' plough (named after the inventor's home town), the first iron plough to achieve significant commercial success. In 1776, James Small introduced another metal plough and this was a true application of science to agriculture, since its design was the outcome of mathematical studies on the shape of the moldboard. Small did not patent it, probably intentionally, with the aim of making it available to as many farmers as possible.

Finally, in 1784, Andrew Meikle introduced a thresher that eventually replaced threshing using a flail, an operation that at the time absorbed almost 25 percent of agricultural work. The device took decades to establish itself, and eventually produced social problems as it caused the unemployment of many agricultural workers.

Alongside these technological innovations came the introduction of new plants (the potato in northern Europe and maize in the Mediterranean area) and a major transformation of land ownership, with farmers, small owners or tenants being replaced by paid agricultural workers who farmed the fields owned by an agrarian owner-entrepreneur.

In traditional societies, more than 75 percent of the working population was employed in agriculture, but by the beginning of the nineteenth century in Britain only one third of the working population worked in the fields. Moreover, this increase in the numbers available to be employed in industry was accompanied by a large increase of the population in general, which doubled in England between 1740 and 1820.

While the agricultural revolution had a noticeable effect on the Industrial Revolution, the reverse effect was only marginal, given that the spread of agricultural machinery only began after 1820.

The increase in population did not go unnoticed. The British economist Thomas Malthus (1766–1834) developed a theory aimed at explaining the impoverishment of the population due to its uncontrolled increase. Until then, it was considered that the population growth was an asset for the state. In contrast, in his '*Essay on the Population Principles*', published in 1798, Malthus argued that while the population grew geometrically, the available resources, and in particular food, grew much more slowly, arithmetically. This meant that the production of resources would not be able to sustain the growth of the population, with a reduction of the resources per capita. As a result, Malthus suggested measures to discourage the birth rate, such as postponing the age of marriage and premarital chastity. In the absence of such measures, it would likely be wars, famines and epidemics that would limit the uncontrolled increase of the population.

Malthus thought, for example, that the welfare policies which had been launched by the British government to improve the conditions of the poor were in fact harmful, because they acted as a stimulus to increase the demographic and could only make things worse by triggering even more serious overpopulation crises.

Malthus' theories have been proposed again several times in the more than 200 years since the publication of the *Essay on the Population Principle*s, in forms that are today called 'Malthusianism' or 'Neo-Malthusianism'. Moreover, while in the original version they referred to agricultural production and to the possibility of feeding a growing population, these criteria were later broadened to refer to the limitations of resources in general and to the subsequent limits to economic development. Since the mid-twentieth century, they have taken on more of an environmentalist leaning and such limitations have been ascribed not only to the impossibility of feeding a growing population, or of finding the resources for unbounded development, but above all to the inability of the ecosystem to withstand the stresses of this growth. More recently, the anthropogenic production of greenhouse gases and the subsequent climate changes have been viewed as the main limiting factors for development.

To support these Malthusian theses, mathematical models which try to describe the dynamics linking economic growth, population and resources are now often used. The forerunner (and the most famous) of these studies was the report '*The limits of development*' written for the Club of Rome in 1972 [1]. Of course, the catastrophic forecasts reported in that study, as well as almost all the Neo-Malthusian forecasts, fortunately did not materialize.

Ever since they were formalized, Malthus's theories were subject to very strong criticism. Karl Marx (1818–1883), for example, denied they had any originality, defining the *Essay on the Population Principle* as "*a superficial plagiarism, the declamation by a scholar, or better by a priest, of the works by Defoe, Sir James Steuart, Townsend, Franklin, Wallace, etc., and does not even contain a single original position*" [2].

More specifically, many have expressed the objection that the progress of science and technology is constantly able to fuel an exponential growth of the population, but we will return to this point in due course.

10.2 THE FIRST INDUSTRIAL REVOLUTION

The term 'Industrial Revolution' refers to those developments in industry which began in the second half of the eighteenth century in Britain and deeply changed the way all types of goods were produced, causing what we could call a transition from craftsmanship to industry. If we want to date this more precisely, it can be said that it took place in the period from 1760 to 1840.

Its main features were the introduction of machine tools, together with a production system based on factories, the increased importance of the chemical and steel industry, and the spread of steam engines, first in industry and then in transportation. However, it would be a mistake to think that the steam engine was the primary energy source of the Industrial Revolution, given that the machines operating in the factories were driven mainly by water wheels, particularly at the beginning.

The core of the Industrial Revolution was the textile industry. The mechanization of cotton spinning led to an increase in productivity by a factor of 500, and that of weaving by a factor of 40. The inventions of spinning machines and looms followed one another quite rapidly, and this, plus the availability of raw cotton coming from American plantations and the increased productivity, allowed the

British industry to overcome its competitive disadvantage with respect to Indian and Chinese cotton of the much lower wages of Eastern workers. Despite the better quality of the hand-spun and woven cotton from India, the reduced production costs allowed by mechanization enabled the British industry to become dominant in the low cost, mass production market.

However, the Industrial Revolution also indirectly encouraged slavery in the Americas, as the strong demand for cotton increased its price, making it extremely convenient to grow in North American and also Brazilian plantations. The American Civil War and the end of slavery encouraged cotton production in other areas, such as the African colonies and Uzbekistan.

Large increases in productivity, although lower than those related to cotton, also occurred in the production of wool and linen fabrics. Even the spinning of silk, which had already been practiced in Italy for centuries using machines of considerable complexity in a climate of great secrecy, spread to Britain and benefited from the general mechanization of the textile industry. John Lombe was hired in Italy in the silk industry and managed to copy its secrets, then opened a factory in Derby which was successful technically, but encountered problems due to the ban on the export of raw silk from Italy and had to receive help directly from the Crown.

One of the most significant features of the Industrial Revolution was the rise to prominence of the steel industry. Since the end of the Middle Ages, coal had been known and used in small quantities as fuel, but its use only began to spread from 1600, when increased demand for wood, as a building material, for the production of charcoal and also to smelt iron and other metals, led to its scarcity. In 1688, it was discovered that a hard and resistant residue, together with a mixture of combustible gases, could be obtained by heating coal at high temperature.

This residue was called coke and it gradually replaced charcoal in iron smelting, enabling better quality iron to be produced more cheaply. In 1730, the first blast furnaces were built in Britain. Operating at temperatures above 1,150°C, they produced molten cast iron, although the refining of cast iron to obtain steel was a problem that kept industrialists and inventors busy for more than a century. A partial solution occurred at the end of the eighteenth century with the introduction of puddled iron. This was followed in the mid-nineteenth century by the construction of converters, which made it possible to obtain steel as we know it today.

In addition, the gases produced along with the coke, if collected and purified, burned with very bright light. They were called 'lighting gases' and at the beginning of the nineteenth century there were distribution networks for lighting the streets and houses. The first public street gas lighting was carried out in Pall Mall, London, in January 1807. Gas lighting, with all its importance in making days longer both for working and for life in general, can therefore be considered as a by-product of the steel industry which developed during the Industrial Revolution.

10.2 The First Industrial Revolution

Iron (cast iron and steel) became widespread in the construction of machines of all kinds, gradually replacing wooden structures. While machines could be made entirely of wood up till about 1600, steam engines required at least some metal parts. It was only the possibility of producing good quality iron at low cost, therefore, which allowed the construction of the new machines of the Industrial Revolution – although some the most advanced machines of the past, such as the hydraulic pumps of the Roman era, had necessarily already been made of metal.

However, the gradual replacement of wood with iron in traditional machines, such as water wheels and many parts of the structures of wagons, permitted lighter and better machines to be built and also soon reduced their cost. Machine tools, another product of the Industrial Revolution, were almost completely made of metal. In general, machine tools did not have their own motor, but they were powered by the drive machine which supplied the required power to the entire factory through a system of shafts, pulleys and belts, (Fig. 10.1). This engine could be a steam engine but, as mentioned, was usually a water wheel, especially early on.

Figure 10.1. Machine tools driven by a single engine, using shafts and belts. [Image from https://upload.wikimedia.org/wikipedia/commons/9/96/Bild_Maschinenhalle_Escher_Wyss_1875.jpg.]

The use of machine tools led to a noteworthy increase in machining precision, which in turn made it possible to improve the performance of all machines, reducing the need for adjustments and enabling the components to be standardized. The effects of this greater precision have lasted well beyond the period considered here and were the basis of the mass production of the twentieth century.

In a short time, iron was also being used in the structures of bridges. The first iron bridge was built by Abraham Darby on the River Severn at Coalbrookdale in Shropshire (Fig. 10.2). It was a very successful project, which was taken as a model for hundreds of bridges built in the nineteenth century. Later, metal parts such as cast-iron columns were also used for residential buildings.

Figure 10.2. The first cast-iron bridge built by Abraham Darby on the River Severn, in Coalbrookdale, Shropshire in 1781. [Image from https://upload.wikimedia.org/wikipedia/commons/f/f0/The_ironbridge_-_geograph.org.uk_-_244380.jpg.]

The Industrial Revolution influenced every aspect of daily life. In traditional societies, the gross domestic product remained essentially constant, with significant exceptions only in periods affected by particular changes, positive or negative, due to exceptional events. With the beginning of the Industrial Revolution a period of rapid development began, both in terms of wealth produced per capita and in terms of population. The consequence was an increase in living standards, although many consider in this latter regard that the effects were only felt much later, at the end of the nineteenth century or even into the twentieth century. Alongside this, however, there was an increase in inequalities and a greater stratification of society.

At the end of the period considered here, that is around 1830–1840, there was an economic recession caused by the slowdown in the introduction of new technologies and by the saturation of the market. Growth only resumed around 1870, with what is normally called the Second Industrial Revolution.

In the period preceding the Industrial Revolution, many peasant families were spinning and weaving clothing both for sale and for their own use, under the direction of traders who supplied them with the raw materials and bought their products. The same thing happened in India, China, and various other regions of Asia where cotton fabrics were produced, and in Europe where mainly wool and linen were worked. The simplest, early spinning machines could be bought by the same farmers and used at home, while those typical of the Industrial Revolution were owned by capitalists who operated actual factories. The workforce consisted mainly of children and unmarried women, who worked 12 to 14 hours a day, six days a week.

The reduction in the number of people employed in agriculture led many people to move to cities, although many factories were located in small towns where it was possible to use running water to power the water wheels. The mass movement of people into the cities came about when the steam engine replaced the water wheel as a source of power. While on the one hand there is no doubt that the Industrial Revolution saw the development of a large middle class, which could enjoy steadily improving living conditions and which constituted that internal market needed to absorb the growing volume of products which industry was continuously turning out, on the other, it is also undoubted that the living conditions of the poorest part of the population declined for at least the first part of the Industrial Revolution.

The Industrial Revolution was also the first period in history in which there was a simultaneous increase in population and per capita income.

In 1750, average life expectancy was 35 years in France, 40 years in England and 45–50 years in the American colonies, mainly due to chronic malnutrition, and this did not increase significantly until the second half of the nineteenth century. Even then, the increase in the average lifespan was less to do with an increase in the length of people's lives and more due to a drastic reduction in child mortality. For example, in London from 1730 to 1749, some 74.5 percent of children did not reach five years of age. That figure had reduced to 31.8 percent by 1810 to 1829.

In cities, the housing conditions of the lower classes were miserable, while sanitary conditions were poor enough to make diseases such as cholera, typhoid fever and in some areas tuberculosis endemic, although in the cities the consequences of famine, such as that which devastated Ireland in the 1840s, were much less severe than in rural areas. While many saw their standards of living decrease in the early years of the Industrial Revolution, it should be noted that standards of living in general prior to this time were very low anyway, and phenomena such as child labor, very long working hours, poor hygiene and widespread malnutrition had always been present even in the wealthiest parts of Europe.

The growing inequalities and very difficult conditions in which a significant portion of the population was forced to live led to the creation of trade unions, mutual aid societies and other organizations committed to improving the living conditions of workers. On the other side of this social divide, there were many enlightened entrepreneurs who endeavored to improve the conditions of workers. Thanks to their commitment, and to the legislative interventions promoted by the unions and the political organizations connected with them – the Labour party in England, for example – the general conditions for the lower classes improved substantially during the nineteenth century.

One phenomenon, of particular importance here for its anti-technological aspect, was that of the Luddites, named after Ned Ludd, the legendary leader of the protests that took place in England between 1811 and 1817. Ludd was allegedly a weaver called Edward Ludlam from Anstey, near Leicester, who in 1779 sabotaged two stocking frames. This was a movement that identified mechanization as the reason for the impoverishment of workers and argued that a legislation completely banning the use of machines in industry should be adopted. Given the impossibility of obtaining such laws, the members of the group became violent.

Luddites were initially workers in the textile industry, in particular the lace and stocking industry of Nottingham, who took action after losing their jobs due to the reduction of labor as a consequence of the increased productivity caused by mechanization, and committed acts of sabotage. Their violent actions spread across other sectors, in particular when mechanization spread throughout agriculture when they began destroying and setting fire to threshers.

After the destruction of many machines and of entire factories, the government reacted by sending the army to protect the factories. Many of the Luddites were either hanged or deported to Australia.

It must be said that the movement remained as extremist as it was when it began, and never asked for better conditions of life for workers or other partial objectives, but merely for a law contemplating the complete destruction of any industrial machinery. It was therefore a movement which identified technology itself as the evil to fight and it attracted many others not directly involved with the problem, particularly some owners of small shops. The anti-technology reaction of the Luddites has repeated itself, albeit less violently, in other contexts when technological innovations have endangered jobs.

The use of coal and the increase in population caused a major increase in air pollution in large British cities such as London (Fig. 10.3). As we can see, however, the phenomenon started almost 150 years before the Industrial Revolution and reached its peak before the end of the nineteenth century.

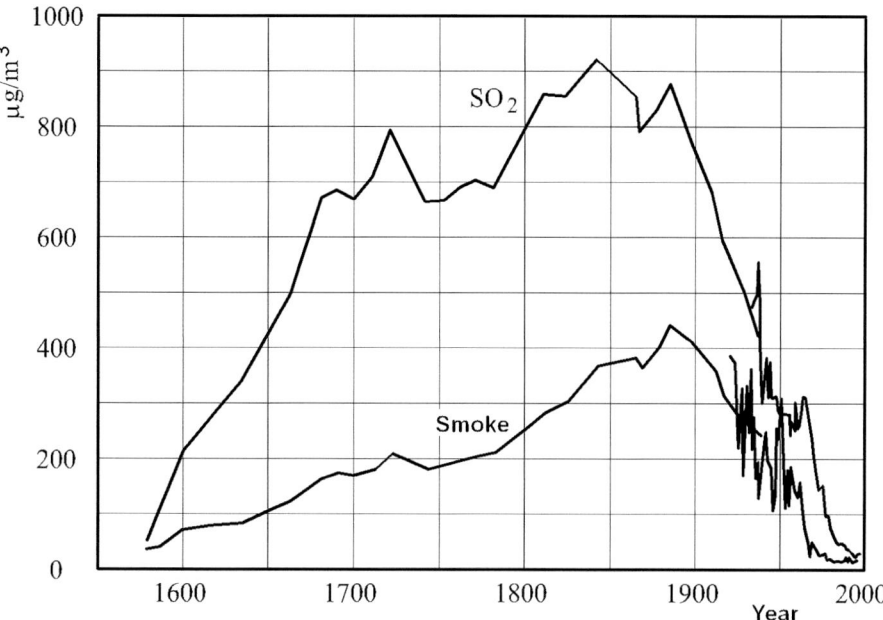

Figure 10.3. Atmospheric pollution, in terms of content of sulfur oxide and particulate matter (smoke) in London air. The data related to the period up to 1930, for which there are no precise measurements, are approximate estimates. [Image from G. Genta, A. Genta, Motor Vehicle Dynamics, Modeling and Simulation, World Scientific, Singapore 2017.]

Similar trends can be identified in such graphs for other cities, although the peaks would be shifted to the right (in other countries, their industrial revolution began later) and downwards (the later the industrial revolution took place, the more precautions were taken to reduce pollution).

The pollution did not just apply to the air, but also to the water and soil. As the various phenomena took place, legislative measures were enacted to contain the pollution, although often, owing to the huge interests at stake, each of these laws was the subject of legal battles and often to quite fierce political clashes.

Most of the above mainly applies to Britain, certainly England, which was the first nation where an industrial revolution occurred. Other countries followed a similar trend, at a later date and often much more quickly, since in many cases the technologies were imported, acquiring patents, hiring British engineers, or hosting local branches of British companies which adapted technologies developed in the motherland to local resources and conditions.

10.3 THE SECOND INDUSTRIAL REVOLUTION

It is generally assumed that the first Industrial Revolution was completed in Britain in the years 1830–1840. By the end of that period, the transition from craftsmanship to industrial production was complete. The steam engine had not only spread throughout industry but also in transportation, in particular with steam ships and railways. Coal had become a universal fuel, both for steam engines and for the production of heat, while iron (cast iron and steel produced with rather primitive processes) had replaced wood in many structural applications.

As noted, there was a period of crisis during those years, of short duration, at the end of which the drive to innovate resumed, albeit with characteristics that over time were different from those typical of the Industrial Revolution. For this reason, what happened in the second half of the nineteenth century and at the beginning of the twentieth is normally called the Second Industrial Revolution. This 'new' industrial revolution began in continuity with the existing one, so that it is difficult to identify precisely when the change took place. Arbitrarily, this could perhaps be identified with the introduction of converters to produce steel from cast iron. The Bessmer converter, which was introduced in 1860 but perfected only in 1870, caused a real revolution.

The 'new' steel performed much better than the puddled iron of the late eighteenth century and the theoretical developments in chemistry and metallurgy allowed hitherto unexpected progress.

Together with the revolution in the steel industry, new alloys were introduced; first aluminum, then magnesium and finally, in the mid-twentieth century, titanium alloys. Composite materials and new synthetic materials, such as thermoplastic and thermosetting resins, slowly entered common use, without replacing steel which maintained its role as the most widespread material.

At the end of the nineteenth century came the electrification and the diffusion of the electrical industry. This revolution, unpredictable only half a century earlier when electrical phenomena were simply a curiosity (like the *mirabila* of ancient times), drastically divided (in space) the production and use of energy, allowing its easy transportation even over great distances.

Workshops, like that shown in Figure 10.1, disappeared. Now, every machine tool had its own electric motor (or rather, several motors for the various 'axes'), while the distribution of energy ceased to be mechanical through shafts and belts and was instead carried out by electric wires. New forms of energy, such as hydroelectric energy from mountain lakes or the great rivers of Canada, could be exploited and transported great distances, and coal-poor regions such as northern Italy were now able to be industrialized rapidly.

A new means of transportation, the automobile, appeared at the end of the nineteenth century. Initially, the motor car was not developed to answer a specific need, because society had perfectly adapted at that point to a transportation system based on railways and canals in areas where the high density of traffic justified the large investments, and on animal-powered vehicles in rural areas.

10.3 The Second Industrial Revolution

In Europe, the automobile was initially more a toy for wealthy gentlemen in search of visibility and new experiences than as a means of transportation. The situation was different in America, where the vast distances and the low density of traffic made the 'horseless carriage' much more useful as a work tool and as a means to ensure widespread mobility. Both a cause and an effect of this difference was the radically opposite fiscal regime to which the automobile and all related products (in particular the fuel for the engines) were subjected on either side of the Atlantic: heavily taxed luxury goods in Europe, and little taxed primary asset in America.

The automobile took several decades to become the dominant means of transportation for the Second Industrial Revolution, but from the outset it was much more than a vehicle. It quickly became a status symbol and gradually became a tool which allowed everyone a freedom of movement that had never previously been possible in history. In particular, thanks to the role of the new mass media, the cinema, it was instrumental in spreading the ideology based on democracy and equal opportunities known as the "American Dream." One notable aspect of this which was closely related to the automobile was freedom of movement for women.

In this regard, it should be noted that in the United States at the beginning of the twentieth century, automobiles were produced in basically three types, in more or less equal quantities: cars powered by internal combustion engines, by steam engines and by electric motors. Starting the former required some physical strength, and, once started, they were noisy and generically 'dirty', more suitable as toys for adults – generally wealthy males – than as a means of transportation for everybody. The steam, and particularly the electric cars on the other hand did not require the use of a hand crank for starting and were silent and 'clean'. They were thus preferred by women drivers. It was only the introduction of the electric starter and the automatic gearbox that permitted cars powered by internal combustion engines to establish themselves on the market beyond any preferences of status and gender.

The motor car was not just a means of transportation, a status symbol and an instrument of freedom, however, it was above all a complex industrial product, to be marketed in large quantities and at the lowest possible cost. These characteristics forced the industry to complete the revolution of production towards mechanization. The assembly line was thus introduced in the automotive industry and it enabled costs to be reduced so substantially that the workers within the industry could afford to buy the cars they built.

Over this same period, the ownership of industries underwent an important evolution. Instead of being companies owned by individual capitalists, they became companies whose shares were more widely owned, by both private individuals and institutions. In many countries, the state entered the shareholding markets at various levels, with companies in sectors deemed as strategic.

With the growing importance of the automobile industry, Britain lost much of its leadership in favor of the United States and, to a lesser extent, Germany. Another consequence was the reduction of the importance of coal as a major source of energy and the growth of the importance of oil.

10.4 THE THIRD INDUSTRIAL REVOLUTION

The Second Industrial Revolution experienced a deep economic crisis triggered by the depression of 1929. It was a lengthy and disastrous crisis, and the economy did not recovery until the outbreak of World War II.

That conflict caused such devastation that the post-war world had to face a radical change in the productive paradigm. Some of the victors, and of course all those who lost the war, had suffered such widespread destruction that they needed years to recover, even though the recovery in Western Europe was accelerated by the crucial assistance of the American Marshall Plan, launched in 1948.

The war effort had triggered a general advancement of technology, not only thanks to the realization of nuclear power systems as a direct consequence of the research for the development of the atomic bomb, the construction of early electronic computers and electronics in general, and the great progress in the aeronautical field, but also due to the introduction of quality control methods in industry. This stemmed from the need to produce large quantities of complex machinery with a much reduced and usually non-specialist workforce, because a significant proportion of the workers employed in the aeronautical and armaments industry had gone off to war. Developed in the United States, these methodologies, mainly based on a scientific approach with the application of statistics to production rather than the products, were then introduced throughout the reconstruction of Japanese industry by W. E. Deming, laying the foundations for what was called the 'Japanese economic miracle'. Thanks to these methodologies, Japan was able to play a key role over the following decades in the development of new technologies such as electronics and mechatronics. We can thus talk of a Third Industrial Revolution, which began with the reconstruction of the post-war period and became fully developed starting from the 1960s.

The automotive industry maintained its dominant role, as cars became more affordable to more people, followed by the aeronautic, and then the aerospace industry. At the end of the war, the new turbojet engine was developed to power aircraft and within in a few years it had replaced the reciprocating piston engine, first in military and then in civilian aircraft. It was an innovation that made long-range aviation safer, more comfortable and above all cheaper, and began the era of mass international tourism.

Perhaps the most important feature of the Third Industrial Revolution was the major expansion of the internal market; so-called 'consumerism'. The ever-increasing rates and efficiency of production of goods and services required a market in which to sell these products, and it therefore became necessary to convince potential consumers to buy them, many of which they probably did not need. In addition, more and more these purchases were becoming available on finance through credit institutions, which meant that even people who did not yet have the money to afford these goods could purchase them anyway.

10.4 The Third Industrial Revolution

The Cold War between the two main powers on the winning side in World War II, the United States and the Soviet Union, caused huge quantities of money to flow into the so-called military-industrial complex of both sides, even though the Cold War did not degenerate into an actual war owing to the fearful destructive power of the nuclear weapons that both sides had available.

From 1957 when, under the auspices of the International Geophysical Year, the Soviet Union placed the first artificial satellite into orbit, what at first appeared to be a long-term commitment to space exploration instead became the arena in which the two superpowers competed, in a race aimed at demonstrating the superiority of their respective economic and political systems.

Only 12 years on from that event, the first pair of American astronauts landed on the Moon (Fig. 10.4a). The effort was immense and had far-reaching consequences for all branches of technology. The quantity of new products which were invented during this effort was enormous, but the most important outcome was the development of new methods of design and structural analysis. After that turning point of technology, no project was ever developed again as it had been before. It has been stated that the investments made for the lunar adventure produced a 500 percent return, just from patents and the economic exploitation of the technological innovations. The lunar adventure triggered a period of economic growth that lasted several decades.

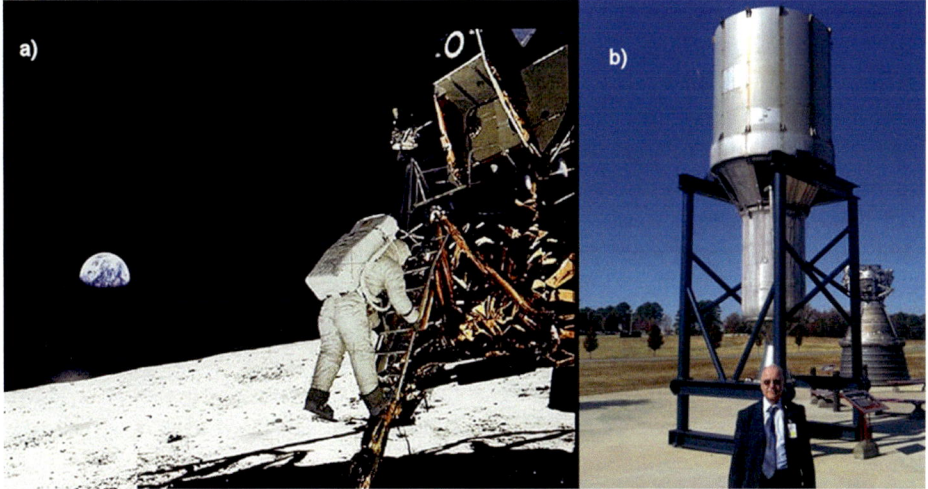

Figure 10.4. a) The second man about to step onto the lunar surface: Buzz Aldrin, photographed by Neil Armstrong who had reached the surface a few minutes earlier. [Image courtesy NASA.] b) The nuclear engine of the NERVA project (Nuclear Engine for Rocket Vehicle Application) exhibited at the museum in Huntsville, Alabama. One of the authors of this book is depicted in the foreground, to give an idea of the small size of the engine, which also includes the nuclear reactor.

But the Apollo missions, the project that placed 12 humans on the lunar surface, was intended to be just the beginning. In the 1961 speech by President John Kennedy which revealed the intention of landing astronauts on the Moon, the development of nuclear thrusters (Fig. 10.4b) for human deep space exploration was also announced, to facilitate the actual exploration, and the subsequent colonization, of the solar system.

The Third Industrial Revolution then suffered a setback, at least partly linked with the American defeat in the Vietnam war. It was not just the defeat itself, a severe blow for a nation that had never known a military defeat in its history, and which was perceived to be the greatest world power. The feeling that it would have been possible to avoid the defeat, and the sympathy that the cause of the 'enemy' received both from a large section of the American public – particularly among intellectuals – and from the allied countries belonging to the North Atlantic Treaty Organization (NATO), left the United States divided and in an unprecedented moral crisis.

In the early 1970s, the United States and its allies were involved in a number of technological and industrial setbacks that came very close to aborting the Third Industrial Revolution. Many Western countries renounced nuclear power, while those who did not took a backward step towards fossil fuels, mainly oil derivatives, at a time of sharp oil price increases which were not justified by a presumed scarcity of oil but purely due to political and speculative reasons. The role of oil-related lobbies in the decisions regarding the retreat from nuclear power that occurred in those years remains to be clarified.

A second setback was the cancellation of deep space exploration programs, in particular those regarding human presence in outer space beyond low Earth orbit and on the Moon. The cancellation of projects such as lunar bases and human missions to Mars, planned for the 1980s and certainly before the end of the century, led to the stagnation of space exploration. The nuclear thrusters, which would have opened the solar system to humankind, had already been tested on the bench and had shown considerable reliability, coming very close to actual testing in space. However, all of the related programs were cancelled, first in the United States (1972) and then in the Soviet Union (with the fall of the USSR) [3]. The cancellation of the nuclear propulsion programs was reminiscent of the destruction of the fleet of Admiral Zheng He centuries earlier by the Chinese government.

A third missed opportunity was the creation of civilian supersonic aircraft, which led to the stagnation of the aeronautical industry – whose progress was limited to details – for decades.

Likely as a consequence of the failure to proceed in the above mentioned directions, the Third Industrial Revolution instead focused on a sector normally called ICT (Information and [Tele]Communications Technologies), which from the beginning had played a supporting role in the development of other industries – mainly aerospace – and now became the focus of innovation itself. ICT played a twofold role in all kinds of industries by greatly influencing the development of both the products and the production processes. In other words, not only did they influence what was produced, but also how to design and how to realize the products. Computers became heavily involved in the design of any artefact and in its production plants, taking control of the machine tools and production processes.

Another important effect of introducing computers into the design of products of all kinds is linked to the so-called three-dimensional CAD (the acronym CAD stands for Computer-Aided Design). Thanks to the graphic potential of modern computers, it is now possible to visualize the object that is being designed directly in a three-dimensional view, without having to pass through the two-dimensional representation of the technical drawing. After two hundred years in which technical drawing had established itself over the three-dimensional representations of the "theaters of machines", the latter has now made a comeback, adding the precision allowed by computers to the intuitive solid representation. It seems that technical design as it has been known by generations of designers is due to disappear and, in the future, computers will make it possible to move directly from the three-dimensional concept to the actual object built by machine tools.

Computers have actually caused a much greater change. Their huge computational power has changed the methods, the mathematical instruments and even the language used to build the mathematical models used in theoretical and applied sciences.

Generally speaking, while it was necessary before the advent of powerful computers to use large scale reductionism, isolating the aspects considered essential and creating extremely simplified mathematical models, today it is possible to solve models of extreme complexity numerically, made up of millions (or several tens of millions) of algebraic or differential equations.

As far back as 1978, Tinsley Oden and Klaus-Jürgen Bathe, speaking of structural analysis, identified what they called a disease, namely "*number-crunching syndrome*", which was affecting those who dealt with computational mechanics. They defined it as the "*blatant overconfidence, indeed the arrogance, of many working in the field that is becoming a disease of epidemic proportions in the computational mechanics community. Acute symptoms are the naïve viewpoint that because gargantuan computers are now available, one can code all the complicated equations of physics, grind out some numbers, and thereby describe every physical phenomena of interest to mankind.*" **[4]**.

These methods and instruments give the user a feeling of omnipotence, because they supply numerical results to problems that can be of astounding complexity without necessarily allowing the user to control the various stages of the computation. This may be particularly dangerous because they can be used by analysts who may be unaware of the subtleties of the methods they are using and above all of the simplifying assumptions which are at the basis of their formulation. Moreover, the users may unquestioningly accept the results obtained from an 'unbiased' and 'infallible' machine.

Generally speaking, a mathematical model is acceptable only if it yields predictions close to the actual behavior of the physical system. The analyst, who is using complex numerical methods, must have an understanding of the physics of the phenomena under study which is scarcely less than that of the specialist who has prepared the code, and must always validate the numerical results by comparing them with experimental results.

The use of sophisticated computational methods must not replace the skill of building very simple models that retain the basic features of the actual system with a minimum of complexity, as required by the reductionist approach. The illusion of approaching a problem in a 'holistic' way, that is, trying to include all the aspects of the actual physical system into the model, may lead to results which are so complex that they are very difficult to interpret and end up by shedding very little light on the behavior of the actual system. This does not mean that we must refrain from using the enormous computational potential that modern computer science puts at our disposal, in particular given that we now know that complex systems (starting from nonlinear mechanical systems and moving up to biological systems of increasing complexity) show 'emergent' properties which are present in the system as a whole and not in its individual parts [5]. What we must do is proceed step-by-step, starting from very simple (usually analytical) models and increasing their complexity to include the various additional aspects until we have a model in which the whole behavior of the system is represented.

Above all, we must always be careful to avoid falling into the mysticism of 'complexity' and 'holism', to reach results which have little to do with the 'scientific' and fall into the realm of 'magical' thought (see section 11.1), as pointed out by Alain Sokal and Jean Bricmont [6]. Additionally, we must always validate the numerical results obtained using complex mathematical models experimentally.

The miniaturization and the constantly increasing power of electronics (see section 12.2) allowed first analogue systems (the so-called PLC, Programmable Logic Controllers), then digital systems (the microprocessors) to be incorporated into even the smallest consumer goods. The first example of mechatronic consumer goods were the cameras produced in the millions by Japanese industry.

The telecommunications revolution that resulted from the development of ICT transformed the world, creating what is normally referred to as 'the global village'. Then it was the turn of the Internet, a technology introduced by the military to allow the commanders to keep in touch constantly and to control the situation, even in the eventuality of the massive devastation caused by a nuclear war. The Internet not only quickly transformed the way people communicate, but also the way people keep themselves informed and circulate news. The world of printed paper thus gave way to electronic information, but while it has made it considerably easier to circulate information of any kind, it has also made it much more difficult to check the reliability of the sources and has enabled the uncontrolled proliferation of fake news.

10.5 INDUSTRIAL REVOLUTION 4.0

We talk of a Fourth Industrial Revolution – or rather of Industrial Revolution 4.0, to indulge in the fashion for assessing the different stages of a phenomenon with the designation used for the various versions of a computer program, which explicitly stresses the importance that ICT plays in this phase of the industrial revolution – to indicate the current phase which started at the turn of the millennium.

It must be noted that talking about the Fourth Industrial Revolution is quite difficult at present, as always when trying to extract the fundamental lines of developments which are currently underway. The proximity of the events makes it very hard to tell apparent characteristics and intellectual fashions from the real trends at the core of the phenomena, and the same cause-effect relationships may be unclear.

We often consider Industrial Revolution 4.0 to be the direct product of ICT and globalization, forgetting aspects that will probably only be considered in their full importance in the future.

Certainly, ICT now allows us to create virtual realities to accompany the actual reality and provides a more detailed understanding of the world of living matter, thus making important progress in science and technology possible. The possibility of dealing with a huge amount of data (the so-called 'Big Data') and of studying complex phenomena has an enormous importance in fields such as biology and medicine, which made huge advances with the decoding of human DNA. The knowledge and treatment of genetic diseases are radically changing, while greater knowledge of the aging processes will also lead to revolutionary progress.

The view that while the twentieth century had been the century of physical sciences, so the twenty-first century will be the century of biological sciences, is much more prevalent. The introduction of genetically modified organisms (GMOs) into agriculture is the modern equivalent of an ancient practice. Agriculture actually began with genetically modified organisms, but the difference lies in how the modifications of the DNA are carried out. Since the Neolithic Revolution, modifying the genome of a plant or of an animal required trial and error by crossbreeding and selecting the appropriate varieties of the species. Without knowing why and how, the genetic heritage of wild species was thoroughly modified to obtain domestic species. One example is the previously mentioned case of the wild almond tree that produced poisonous fruits, which was modified to obtain the domestic almond tree whose fruits are edible.

In the aftermath of World War II, after realizing that radiation produces genetic mutations, large experimental facilities were created in which many plants were placed all around a radioactive source to create random mutations. Then, selecting the plants that produced favorable mutations, new varieties or new species were created. In Italy, at the Center for Nuclear Studies of the CNEN at Casaccia, Rome in 1974, a new variety of hard wheat (variety Cp B144) was thus obtained by bombarding the usual varieties of wheat with X-rays and γ-rays, and then crossing the products of the mutations. The wheat produced is widely used in the production of pasta – or at least of high-quality pasta. Many of the plants currently used in agriculture are the result of radiation-induced mutations during such experiments.

Thanks to the study of DNA, it is now possible to discontinue these empirical practices and actually design the DNA of plant and animal species, modifying their genome in a conscious and rational way rather than 'hammering' it with radiation to induce random mutations and then selecting the favorable ones. Species particularly resistant to drought or to parasites have been created, in order to increase agricultural output without using large quantities of pesticides, or to exploit soils which were previously considered unproductive. Of course, the use of genetically modified organisms is one of the answers to the original theories of Malthus concerning the possibility of feeding a growing population.

As for the other Neo-Malthusian fears, relating to the depletion of resources, it must be said that the reserves of oil and other raw materials actually increase over time rather than decrease. The discoveries of new oil or mineral fields constantly exceed extractions. In fact, the problem is not so much the scarcity of raw materials, but the increasing costs of extraction and refining from increasingly poor deposits. In theory, it would be a mistake to think that there will never be problems in obtaining raw materials. Of course, the planet contains a limited quantity of resources, and sooner or later we will reach those limits to development of which the Neo-Malthusians speak. As we will see, the solution is to start exploiting the resources beyond our planet before the costs of raw materials increase in such a way as to compromise development – something that currently appears some way off in the future.

Among the aspects of the Fourth Industrial Revolution that are not related to ICT is a return to approaches that had been interrupted in the 1970s, at least partially. During the 2010s, there was a lot of discussion about a nuclear renaissance, at least in the specialized journals, to indicate a wide revival of the construction of nuclear power stations. This is a trend that has only marginally touched the West – in quite different ways from nation to nation – and has been concentrated mainly in India and particularly in China. Despite the arguments for the need to combat climate change, the West remains dependent on fossil fuels, particularly oil and natural gas, seemingly in the hope that mythical renewable energies can solve all problems without resorting to the much-demonized nuclear power. In contrast, China is well aware of the fact that only a massive increase in the use of nuclear energy, appropriately mixed with solar and other renewable energies, will enable us to reduce the use of first the more polluting coal and then of oil and natural gas. Only in this way it will be possible to proceed to the electrification of the transportation system, which will require a drastic increase in the production of electric energy.

Another 'comeback', directly from the 1970s, is the development of commercial supersonic aircraft, this time mainly by private companies. In fact, today there are several options, from supersonic atmospheric planes, to faster ballistic vehicles that leave the atmosphere and return at hypersonic speed.

Finally, the most important trend is a renewal of space exploration, but this time stressing its commercial and private aspects rather than the government-backed approach linked to national prestige, even if the renewal of the competition of the 1960s can be foreseen, this time involving other players such as China and India, as well as Korea and Japan.

The most significant aspect, however, is the major presence of private individuals and companies, who are usually referred to using the term 'New Space'. This approach has already produced a significant reduction in launch costs and the creation of a new generation of reusable launch vehicles not directly derived from military ballistic missiles. The business models of New Space are varied: from space tourism to the extraction of mineral resources from the asteroids; and from the transportation of astronauts and equipment from Earth to the International Space Station and then to the Moon and possibly to Mars, to performing scientific experiments and prospecting missions on behalf of 'clients' such as space agencies. This model, born in the United States but now expanding to countries with mixed economies (such as Chinese Market Socialism), is based on the creation of an actual space economy, including the Moon first, then the asteroids and Mars.

A space economy based on the use of terrestrial orbits (low orbits, but above all the geostationary orbit), for the telecommunications market but also for the collection of information (meteorological, related to the use of resources, etc.) and for navigation, has been flourishing for at least 30 years, and its global volume has exceeded the expenses of all the space agencies for more than a decade. But here we are talking about an economy that deals with deep space, up to cislunar space and beyond.

Finally, when speaking of Industrial Revolution 4.0, one cannot overlook that which is often associated with the term 'Industry 4.0', namely the utilization of ICTs throughout, not only in services but above all in the manufacturing industry, from the micro- and nano-components of machines up to the building industry. It is not just a continuation of the trend that began with Computer Numerical Control (CNC) machines, in which conventional machine tools such as lathes and in particular multi-axes milling machines were placed under the control of digital computers, but of something radically new and revolutionary compared to what humans have done since the Palaeolithic.

For instance, in the Paleolithic when a man created an arrow point, he took a suitable pebble and, with great skill, removed all the material which was not needed to obtain the desired artefact, like a sculptor who 'frees' his statue from the block of marble which contains it, or like the turner who removes the chips from the raw material to derive the part described on the drawing. In technical jargon, this kind of operation is called 'machining by chip removal' and had changed little over the last two million years, roughly up to the year 2000. Even the numerical control machines of the Third Industrial Revolution did the same, but simply under the control of a computer instead of a human, although a human still had to program the computer.

In the 1990s, a radically different idea began. Instead of working from a block of material and removing chips, we could start from the very atoms and molecules, assembling them appropriately to create the desired artefact. This process, called 'bottom-up manufacturing', is no different to the approach followed by nature to build living structures, including ourselves. Obedient to the program contained in the DNA, enzymes assemble the amino acids, then the proteins, and finally the cells to realize the tissues of plant and animal organisms. By imitating this process, nanotechnologies promised to create objects of any kind, up to machines of great complexity such as robots or spaceships [7]. However, this way of working to construct artefacts from the 'bottom-up' on a microscopic scale had not been realized in the predicted time, namely by the end of the twentieth century.

Instead, a similar way of proceeding has been realized at the macroscopic level. Instead of starting from atoms and molecules, the process starts from powders or thin filaments, which are sintered or melted under the control of a computer, until artifacts of great complexity are built in a simple and economical way. This process is called 'additive manufacturing' or 3-D printing, with the first term generally used for industrial technologies using metallic alloys and the second for processes where simple three-dimensional printers are controlled by a personal computer to create plastic elements, often at the hobby level.

The bottom-up manufacturing processes on a macroscopic scale have begun to revolutionize fields in which small-scale production is common, such as in the aerospace sector. This has reduced costs (sometimes by orders of magnitude) and development times while simultaneously increasing performance thanks to the possibility of building parts of a complexity previously unimaginable. Currently, there are already mass-produced elements available, such as prostheses and some complex parts for the automobile industry, and it is expected that such technologies will become widespread even in large-scale production.

Thanks to the scalability of this process, huge three-dimensional printers have already built homes starting from simple materials such as sand or gravel, while at the other end of the dimensional scale, microscopic printers can produce micro-robots or nano-robots at very low cost which can, for example, operate inside the human body to perform surgical operations without making cuts.

Additive manufacturing technologies are so recent that we have only just begun to understand the new applications which will be made possible, but it is already possible to realize that the field where they will have the greatest impact is the space economy. It could quickly be possible to make machinery and dwellings directly on the Moon or on Mars using local materials, launching from Earth only what is needed to start the process.

At present (well into the Fourth Industrial Revolution) the Earth's technosphere has a total mass of 30 billion, billion tons, that is, 50 kg for every square meter of the Earth's surface.

10.6 SOME CONSIDERATIONS ON INDUSTRIAL REVOLUTIONS

In the previous chapters, it has been noted several times that the revolution represented by scientific technology began in Europe. The same point can be made about the Industrial Revolution. While many of the inventions that have spread since the late Middle Ages, such as the press, the gunpowder and the compass, originated in China, we have to ask the question why did they become common only in Europe?

The answer is once again the same. Only in Europe was there a coming together of factors, ranging from the absolute trust in human rationality (to which a meaning that transcended the human sphere and connected directly with the divine was attributed) and therefore granting the possibility to develop a science that could be applied to technology, to the respect for private property, an essential element in convincing those who were sufficiently wealthy not to consider their wealth as a means to increase their level of consumption but as a tool to generate further wealth by investing it.

Rodney Stark points out that Ali Pascia, the Turkish commander of the fleet which fought at Lepanto against the Christian Fleet, loaded all his belongings on his ship because he obviously could not risk leaving them at home and had no way to invest them. A society in which the wealthy must behave in that way will never have an industrial revolution. Stark also reports that when the steel industry began a development that could have been revolutionary in China, the emperor chose to nationalize it, causing it to collapse. The certainty of private property was at least as important as scientific technology for the birth of the Industrial Revolution and the beginning of capitalism.

The Industrial Revolution began in Europe, and in Britain in particular, because Britain had brought a culture based on freedom, private property and rationality to the highest level. The incubation was very long and began in the much-maligned Middle Ages.

This cultural peculiarity of the West (it is necessary at this point to speak of Western culture and not of European culture, since North America must also be included) was completely understood by the Europeans themselves. Imperial Britain took upon itself a mission to civilize – and to comply with this Britons had a duty to sacrifice even their lives, if necessary – aimed at bringing the benefits stemming from its lifestyle and from its own economic and political system to all peoples. From the mid-eighteenth century, and thus in connection with the beginning of the Industrial Revolution, this vision permeated the works of writers such as Daniel Defoe and was formalized by Rudyard Kipling as the '*Burden of the white man*', to use the title of his famous poem.

The seemingly racist connotation of that very sentence, 'burden of the white man', was in fact linked to Herbert Spencer's social Darwinism, which was then widespread, but was probably not an essential feature. These racial implications lost importance and were replaced by cultural aspects, and the British Empire ended having acquired that multiracial aspect of all the empires throughout history, in which the peoples that had become part of it were assimilated through a cultural integration, regardless of their racial origin.

The elite of the peoples who became part of the British Empire, even those who eventually fought to secede from the Empire (such as in India), and those of the countries that opposed the empire (China, Turkey, etc.), were engaged in the advancement of their countries which was at the same time both modernization and Westernization. This effect is even more remarkable in those countries which, after decolonization and during the Cold War, opted for Soviet-style socialist or communist regimes. After all, Marxism, to which all these countries aspired with their economic systems, is a typically Western philosophy and is drastically opposed to traditional cultures.

As Samuel P. Huntington acutely (and prophetically, since at the time he wrote his book these phenomena were only at the beginning) noted in *The Clash of Civilizations*, many of those countries that initially accepted Westernization as a part of modernization are now rejecting it, with a rebirth of traditional cultures. Very obvious examples are Turkey, with the weakening of Kemalism and the Islamic revival which is a general constant in almost all Islamic countries, but probably also China and India. It remains to be seen how well modernization will survive the repudiation of the culture which invented and promoted it.

Moreover, Europe itself and the United States are also going through a period in which the rejection of Western culture, of Greek-Judeo-Christian origin, is becoming stronger under the pressure of a cultural relativism which, while being in some way a child of the extreme individualism originated by the Enlightenment, is leading them towards a more widespread multiculturalism. In the coming decades, we will see whether the Industrial Revolution, and the high standards of living which followed, will survive the loss of its cultural bases.

References

1. Meadows D.H., Meadows D.L., Randers J., Behrens W.W., *Limits to Growth*, 1972.
2. Marx, K., *The Capital*, 1867
3. Dewar J., *The Nuclear Rocket*, Apogee Books, Burlington, Canada 2009.
4. Oden T.J., Bathe K.J., *A Commentary on Computational Mechanics*, Applied Mechanics Review, 31, p.1053, 1978.
5. Morowitz H.J., *The emergence of everything*, Oxford University Press, Oxford 2002.
6. Sokal A., Bricmont J., *Intellectual Impostures,* Picador, London 1999, introduction.
7. Drexler K.E., *Engines of Creation, The Coming Era of Nanotechnology*, Anchor Books, New York 1986.
8. Huntington S.P., *The clash of civilizations and the remaking of world order,* Simon & Schuster, New York 1996.

11

The irrationalistic constant

11.1　AT THE ORIGINS OF MAGICAL THOUGHT

The history we have related so far has attempted to summarize the journey of *homo sapiens* from prehistory to the beginning of the conquest of space, paying particular attention to the events that saw Western civilization take on the role of protagonist. The focal point of this path is the increasing role of the logical-rational component of humankind, which has been at the center of hopes and expectations over the future of our species since the Middle Ages. We have spoken, in this regard, of an authentic "faith in progress" that over the centuries, without interruption, has motivated millenarists, monks, merchants, capitalists and scientists. As we said at the beginning, humans are technological animals, and technology as we know it today is a product of rationality. Many innovations, from money to the wheel and from the airplane to nuclear energy, have been realized to answer to specific human needs, starting from a logical elaboration of the underlying problems.

However, it would be seriously wrong to reduce *homo sapiens* to a creature of pure intellect. The history traced so far does not regard the human species as a whole, but only looks to its rationality. To offer a complete picture, we must also take into account its opposite dimension, which finds its expression in so-called 'magical thought'.

Psychoanalysts talk about a cognitive process in which the logical relationship between the subject and the object is missing. Such absence is not due to a mistake, but is an intentional refusal of logical structure, which is considered to be an unduly strict limitation on a full understanding of reality. The belief that part of human existence escapes the laws of reason, or indeed does not require (and is not subjected to) any rational demonstration, probably dates back to about 50,000 years ago, and is intimately connected to the cognitive revolution discussed in Section 1.3. In other words, the magical-irrational sphere is as old as rationality itself.

With the development of abstract thought, humans began to think about elements which were beyond their direct sensorial experience and to attribute an existence to intangible entities, as a result of their mental elaboration. From this came the ability to analyze, plan and transform nature which would also give rise to the technological development of *homo sapiens*.

At the same time, however, humans also developed a supernatural sphere as an additional basis for the world surrounding them, and were able to explain its workings as an alternative to direct sensorial experience. When the results of the primary analysis of sensorial data were considered insufficient, magical thought was used.

The most significant cause for such an approach was death, an unresolved problem which, more than any other, stimulates irrational thought. Unsurprisingly, the first elaborations by *homo sapiens* following the cognitive revolution were the funerary practices to facilitate the journey of the deceased in the underworld.

At the same time, the first form of philosophical-rational elaboration on existence was also derived from death. In the face of a complex problem, humans always elaborate in two ways; one of a strictly logical nature, and the other magical-irrational. Between these two extremes we find religion, which since the Neolithic age has represented a meeting ground between magic and rational investigation.

Gradually, the sphere of action of magical thought was extended to the whole natural world and to its innumerable inexplicable elements. Whenever there was a lack of a clear relationship between subject and object, the associative process was used, i.e., the construction of an analogous coupling between similar elements, even where this relationship could not be demonstrated. For many primitive civilizations, fire, air, water, earth and the other forces of nature thus became animate entities endowed with their own will which belonged to the supernatural sphere.

This is not, however, a phenomenon limited to prehistory. Throughout the ages, a large number of different civilizations located around the planet related themselves with the world around them by relying on magical and irrational beliefs of varying complexity and sophistication. This process is not merely a makeshift solution in lieu of a rational scientific explanation, but is perceived as a necessity. According to almost all human cultures, nature itself cannot be explained rationally, as it is governed by cosmic laws that have nothing in common with the human mind. The existence of a relationship between the rational thought of humans and the general order of the universe is a characteristic exclusive to Greek philosophy and the Judeo-Christian religion, and is rejected in a particularly radical way by all Eastern civilizations.

By way of explanation, rational thought makes systematic use of demonstrations. The Greek philosophers were the first to study this process, codifying the main standard steps such as syllogisms. This signified the birth of 'logics', which in Greek literally means 'the art of the word'. In fact, these basic rules should govern every discourse and every scientific reasoning.

In contrast, irrational associative thought proceeds by overdetermination. Each image is followed by another, without giving particular importance to linearity and the absence of repetitions. The same concept is often repeated several times, and each time a new element is added which introduces further new elements and constitutes an enrichment with respect to what has been said previously.

This approach to explanation is typical of the Eastern world, which finds ample expression in the wisdom maxims of India, China and the Middle East, and has a significant influence, albeit in a mediated form, on some passages of the Jewish Bible.

An emblematic example of this way of writing is represented by the *Upanishads*, the short teachings that condense the intellectual heritage of ancient India (eighth to fourth century BC). The content of these maxims, while unquestionably rich and deep, is never a logical elaboration comparable to the works of Greek philosophers, since they do not pursue the construction of a complex, coherent and rigorously demonstrated argumentative path. The expressive immediacy of the single image is associated with the following one merely as a juxtaposition.

In the West, magical thought first found widespread success through the Greco-Roman mystery cults, linked to the god of intoxication, excess and irrationality, Dionysus, as an alternative to the institutional Olympic religion. Subsequently, this approach inspired both hermeticism, a thousand-year-old esoteric movement inspired by the legendary teachings of Hermes Trismegistus, and above all medieval alchemy. The latter discipline, superficially considered by many to be a kind of rudimentary chemistry, has nothing to do with contemporary science, since it does not study the world of inorganic matter analytically.

For the alchemist, the minerals and their transformations were nothing more than symbolic means to reach a goal totally alien to the scientific world. The goal, in fact, was not to achieve a better understanding of the substances and their transformations with which the alchemist operated, but for the inner improvement of the magician who, working in the alchemical furnace (*athanor*), carried out a path of meditation to lead them to transform their soul from "raw stone" to "gold".

If chemists confine themselves to examining the workings of inorganic nature, then the alchemist was instead the exclusive protagonist of his own investigation, and used these substances to construct a complex universe of symbols that combined stones, gems, stars and constellations to express esoteric concepts of a spiritual nature. Being a magical discipline, alchemy was completely based on analogical association[1]. In medieval magical writings, elements such as lead, quicksilver, sulfur and gold, as well as the planets such as Jupiter, Mars, Mercury

---

[1] In his *A History of Alchemy – The Ancient Science of Changing Common Metals Into Gold*, Tower Publishing, New York City, 1962, the historian Serge Hutin stated that analogy, "*a key notion of alchemy*", was based on "*a strict correspondence between the world as a whole and what operates in the alchemical laboratory. Parallelism that works also on other planes (...): it is not a case, for example, that alchemists are oriented towards the total respect of the rhythms of the Great Book of Nature: the seasons, the planetary configurations, the terrestrial, lunar and solar magnetism...*".

and Saturn, lose all their physical-scientific implications to become symbols of metaphysical processes that take place in the human soul.

Ultimately, alchemy had no interest in better understanding chemical processes to apply them practically to reality, but instead limited itself to structuring the transformations of matter stereotypically, without admitting any innovation or improvement. The trustworthiness of these teachings was based entirely on their antiquity and immutability, and any innovation would have destroyed the network of esoteric symbols and underlying metaphoric associations. Even after the accidental discovery of new chemical reactions, the alchemist would likely have repudiated these novelties.

Chronologically, deriving an evolutionary path between alchemy and chemistry is incorrect, as the two disciplines crossed the West over the centuries as two parallel tracks. Isaac Newton himself was an alchemist, as were many other famous modern well-known scholars. Instead of yielding to science, alchemical magic has recently experienced a particularly intense development in the twentieth century, through works ascribed to Fulcanelli, and has found widespread circulation in the contemporary age, at least at the level of a 'cultural fashion' linked to the New Age movement.

In fact, over the last century, irrationalism has not given way at all to scientific rationality, but instead continues to flank it as that parallel track. At a popular level, it can be seen in the simultaneous spread of New Age thinking, of new sectarian phenomena such as Scientology and in the many occultist trends that reference the figure of Aleister Crowley.

The risk, if anything, is that the two parallel tracks could come close enough to be confused with each other. In 1965, the religious historian Mircea Eliade tried, with foresight, to understand the reasons behind the surprising success of the French scientific magazine *Planète*, which boasted 80,000 subscribers and 100,000 readers despite its high cover price. The magazine, which was considered an actual cultural fashion in the 1960s in open contrast to the dull materialism of the followers of Jean-Paul Sartre and Albert Camus, presented itself as "*an amazing mixture of popular science, occultism, astrology, science fiction and spiritual practices*" re-purposed in a scientific environment. Thanks also to the presence of a scientific committee, Eliade observed that the reader "*is convinced that he is supplied with facts, or at least of responsible hypotheses; he is convinced that he is not, at any rate, fooled*" [1]. Actually, it is clear that most of the contents of the magazine were just esoteric suggestions without any foundation or scientific validity.

All this shows that, for 50,000 years, 'magical thinking' has continued to represent an alternative to the scientific discoveries of the Western world. It is an ancient and radical conflict that, in the twenty-first century, could produce severe consequences for the future of technological innovation.

11.2 IRRATIONALISM AND RELIGION

Irrationalism has also had a strong influence in the religious sphere. A constant of oriental thought is that it is impossible for human intellect to describe divinity, to represent its features or to understand its creation. Because it does not respond to the laws of reason, divinity cannot even be the object of thought.

Consequently, as previously mentioned, the development of theology, of a logical-philosophical investigation of God, is an exclusively Judeo-Christian approach, which has no comparison in the religions of India and China and particularly in Islam. During the same centuries in which even the working of the Trinity was subjected to the investigations of intellectuals in the West, who animatedly discussed whether the Holy Spirit emanated from the Son and the Father or only the latter, Islam dogmatically blocked every reasoning on God with the *bila kayf* rule: "*believe without asking why*". The believer has to accept their state of Islam; that is, complete submission to a God who cannot be understood, explored or represented.

In this kind of religiosity, the transcendent principle – which according to some religions of the far East is so unknowable as to coincide with nothingness – can only be described in the negative, affirming what God is not: "non-mortal", "uncreated", "non-limited", "non-mutable", and so on. At the technical level, this is known as an *apophatic* concept of the knowledge of divinity. In Greek, 'apophasis' means 'negation', and the derivation of the term is not accidental.

In fact, this thinking had a deep influence on Greek philosophy. According to Parmenides (515 BC – 541 BC), the absolute Being was beyond any kind of meaning and understanding, and could be defined only by its perfect existence.

In the first centuries after Christ, the pagan philosopher Plotinus (204–270), leader of the Neo-Platonic movement, maintained that the universe could be explained exclusively in terms of proximity to, and distance from, God, an unknowable and indescribable principle from which everything emanated passively.

With regard to science and the exploration of the world around us, the consequence was plain to see. According to these philosophers, the reality in which we live was the polar opposite, a corrupted and degraded version of the divine realm in which everything was still and perfect. Death, as well as all geological and biological changes that governed the earthly world, were an unfortunate consequence of its distance from God, and were therefore unworthy of the attention of human beings, who had to devote their intellect to concentrating on the immutable and perfect divinity that was at the origin of everything. This provided a dilemma: humans should not turn their rationality to the world around them, because it was unworthy of their attention, but they could also not use it to understand God, who defied all human logic.

The only kind of intellectual activity possible, for Plotinus and for a part of Christianity, was therefore the oriental practice of meditation, a magical-mystical journey towards this unknown source, undertaken by switching off the senses. Starting from this principle, the disciples of Plotinus spread their mystical meditation and the magical discipline of theurgy throughout the Mediterranean, consisting of a sequence of esoteric rituals aimed at entering into communion with the divinity.

Also particularly close to this way of thinking was Gnosticism, an esoteric strand of Christianity which, in the first four centuries, produced the apocryphal gospels of Nag Hammadi and other writings such as the *Gospel of Judas*. Completely refusing to give any attention to the earthly world, which was actually conceived as a cosmic prison from which we must try to escape, these sects believed that the human intellect should be used exclusively for the activities of meditation to allow them to reach the *gnosis*, the secret knowledge of divinity that could not be achieved with logic and reason. Just as in the case of Plotinus, this type of concept led to magic. In the Gnostic gospels of Nag Hammadi, there was no shortage of references to ritual formulas, and this sect also produced the treatise on magics entitled *Books of Jeu*.

Even among the medieval Christians, there were also irrationalist mystics who supported "negative theology". In the end, however, the thinking of Bonaventura di Bagnoregio (1221–1274) prevailed, and in his *Itinerarium Mentis in Deo*, he argued that man could explore God by resorting to both rationality and faith. The scientific disciplines and humanistic studies thus became a fundamental step on the path to the knowledge of God. Without this, Western science would probably not have had the extraordinary flowering that we are now witnessing.

A further contribution, a few years later, came from one of the greatest thinkers of Western history, Thomas Aquinas (1225–1274). Although radically different from the theses of Bonaventure, the thinking of the most famous theologian of the thirteenth century Scholastics still maintained that the study of the earthly world would allow humans to prepare themselves for their encounter with the divine, just as Anselm of Aosta (1033–1109) had already affirmed two centuries before. Without this decisive turning point, we would probably know a very different Christianity and West today.

11.3 THE RETURN OF OUROBOROS

Another characteristic element of irrational magic thought is the cyclic perception of time. It has already been observed that in a perfectly cyclic cosmos governed by Ouroboros – the mythological snake who moves in circles around itself devouring its own tail – no form of innovation and progress can be possible. Everything repeats strictly in the same way and there is *"nothing new under the sun"*.

In antiquity, this thinking, typical of Far Eastern philosophies, was also widespread in the Greek world. Cosmically speaking, the Myth of the Ages and the Stoic philosophy of the Hellenistic age proposed a constant cycle of history, bound to repeat itself to infinity. On an individual level, Pythagoras, Empedocles and Plato supported the existence of metempsychosis, that is, a cycle of *post-mortem* reincarnation of the human soul into a new body.

Later, Judeo-Christian culture proposed the linear concept of history and human life that characterizes the contemporary Western view of the world and makes it possible to talk about progress. Even here, however, there was a significant exception. In the apocryphal gospels of Nag Hammadi, the Gnostics accepted the Eastern idea of a reincarnation cycle, once again proving to be the magical-irrational counterpart of Christianity.

Metempsychosis also implies a deeply negative perception of the earthly world, with the cycle of reincarnation often outlined as a form of sad atonement from which man must try to escape to reach the divine dimension. Buddhists call this *Nirvana* and the Gnostics named it *Pleroma*. Such an outlook implies that it is useless to waste time trying to improve one's existence in this world.

But why are we dealing with these issues today? It is because this is not an exotic phenomenon, linked to a remote past, which has had no bearing upon Western culture for more than a millennium, but is something which can be encountered in the present day.

In the last two centuries, these theories have made a comeback, regaining strength through a magical-esoteric environment and fascinating an ever-widening audience, while also portraying a self-appointed scientific authority. The basis of this revival was the famous German esotericist Rudolf Steiner (1861–1925), the founder of the so-called Anthroposophy. This is a sort of 'occult science' which interweaves some contemporary study areas with the idea that humans can access a higher state of consciousness by coming into contact with the "etheric forces" of the spiritual world.

Steiner's theses, endorsed by a great number of followers, brought reincarnation and karma back into fashion, in part due to the widespread fascination for the Far East at that time. According to anthroposophical doctrines, after death humans would enter the *"kamaloka"* stage. This phase, which could last up to twenty-five years, would allow the individual to separate from his *"etheric body"* and reach the state of a pure spiritual seed, the self. After a further interval ranging from five hundred to one thousand years, the self would receive a new etheric body, choose its parents and reincarnate into a physical form, relinquishing the memory of every previous experience. This idea is also a cycle from which humans are destined to escape. In the beginning, humankind would have had a single collective soul, to which it will finally be able to return once the Darwinian evolution of the species has reached its culmination.

Every human being, according to Steiner, is influenced by previous experiences at both the character level and the medical level, so that the predisposition to infectious diseases, to pneumo-bronchial pathologies and premature aging, for example, is derived from the previous behavior.

Although devoid of any scientific basis, the Steiner method has persisted and won many supporters, and still holds the status of 'complementary medicine'. According to this pseudo-science, vaccines and antibiotics should be refused. From a therapeutic point of view, anthroposophic medicine combines remedies of a homeopathic nature[2] with strictly ritual aspects, operating as a frontier pseudo-science and contributing to the diffusion of dangerous contemporary anti-vaccinations trends. This is a return to myth and magical superstition which, even in this area, is constantly increasing.

Parallel to anthroposophical esoterism, there are some academically unrecognized trends in psychiatry that have endeavored to demonstrate the existence of reincarnation. This is true of the studies by Ian Stevenson (1918–2007) who, from 1960, dedicated himself to investigating over three thousand cases of children who claimed to remember their previous lives. These studies suggested the existence of a link between the existing diseases and phobias and the previous cycles of the individual. In 1997, with his monumental *Reincarnation and Biology: A Contribution to the Etiology of Signs and Congenital Defects*, Stevenson argued that 35 percent of children who claimed to have memory of their previous cycles had unusual physical abnormalities or congenital defects that coincided with the fatal wound which caused their previous death. This suggests that the memory would then be due to the trauma of violent death, which is not completely reconcilable with reincarnation.

His theses, currently supported by his former student Jim B. Tucker, have been repudiated by the entire scientific community as they lack any experimental demonstration beyond a mere review of empirical cases. Nevertheless, they continue to inspire a widespread trend in contemporary pop culture. From the movies of the Wachowski sisters to New Age-inspired music, notions like 'karma' and 'metempsychosis' are experiencing growing influence in the collective imagination.

---

[2] Homeopathy is a method of treatment formulated by the German physician Samuel Hahnemann (1755–1843) in his *Organon of rational medicine* of 1810. This discipline, however, has nothing to do with the scientific universe, as it is based on mere associations which defy scientific proof, such as the principle of infinitesimal dilutions. The risks of the dangerous mixture between the rational scientific and the merely associative approach reaches its peak in homeopathy. See, for instance D. Vione, "*Why homeopathy is not a science*", www.cicap.it, 2004; AA.VV., "*The homéopathie au banc d'essai*" in "La Recherche", 310, 1998, pp. 57-87; and J.M. Abgrall, "*The charlatans of health*", Editori Riuniti, Rome 1999.

Apparently, for metempsychosis as well as for alternative medicine, we can disregard the need for a scientific demonstration even in the twenty-first century, replacing it with simple suggestion deriving from the combination of a certain number of significant cases.

Once again, it appears that anti-scientific magical thought and analogical association are still enjoying excellent health.

11.4 THE SHADOWS OF THE ENLIGHTENMENT

"When you do not believe in God, you are ready to believe anything". This saying by François-René Chateaubriand (1768–1848), often erroneously attributed to Gilbert Keith Chesterton (1874–1936), is perhaps the best possible explanation for the surprising return of magical thought that occurred in nineteenth-century Europe after centuries of scientific rationalism. That return, strange as it may seem, was in reality a direct consequence of the Age of Enlightenment.

In the previous sections, so-called magical thought has always been associated with beliefs of a spiritual-religious nature, whether they were Greek mystery cults, oriental philosophies or Western esoteric doctrines. A strong irrational soul, moreover, is also present within Christianity itself, and manifests itself in the Gnosticism of the apocryphal gospels, in medieval mysticism and in the forms of popular religiosity which, even today, often go beyond the bounds of superstition. Such phenomena, however, had always been limited to a minority and heterodox position.

However, the nineteenth century witnessed the rapid flowering of a large number of doctrines and magical rites in an environment that was apparently rationalist and increasingly dominated by skepticism towards traditional religiosity. Through Allan Kardec (1804–1869), Eliphas Lévi (1810–1875) and Madame Blavatsky (1831–1891), disciplines such as spiritualism, occultism and demonology entered the salons of the upper bourgeoisie, seduced the intellectual elite and influenced a world that, less than a century before, seemed to have banned every form of spirituality while reaching the apex of rationality.

This apparent contradiction was made even more evident by the triumphal tone of the declarations of principle with which the eighteenth century philosophers had celebrated the triumph of rationality, extolling the Enlightenment as the final *"exit of humankind from the state of minority of which it has the final responsibility"* [2].

This "state of minority" was represented, according to many thinkers of the time, by the subjection to a religious faith that for millennia had limited the full expression of human reason. The attempt to dispose of this alleged legacy of the past, however, ended up feeding a new need for spirituality, which manifested itself in the proliferation of magical superstitions and esoteric doctrines of all sorts. To understand this new era of magical thought, therefore, we must look at the many contradictions that permeated the eighteenth-century Enlightenment.

According to Giancarlo Elia Valori, the Enlightenment consisted of various different concepts: "*A typical trait of Enlightenment is the oscillation between deism and atheism, where deism always refers to a God who does not care for the activities of humans, but nevertheless remunerates them according to their merits. Enlightenment atheism, on the other hand, is characterized by hatred towards all revealed religions and by the idea that man and nature can be explained only scientifically and in a completely materialistic way*" [3].

Among the deists[3] can be found, for instance, the philosophers Charles-Louis Montesquieu and Jean-Jacques Rousseau, as well as those Jacobins who introduced the cult of the Goddess Reason and of the Supreme Being who, during the French Revolution, tried to completely redirect Western religiosity by replacing Christianity with a set of sacred rites addressed to this cold spiritual entity. Starting from 1792, after a harsh persecutory campaign against the clergy, many French churches were deconsecrated and dedicated to the cult of the Goddess Reason. In Paris in particular, this phenomenon reached several places of high symbolic value, such as Saint Sulpice, Saint Denis, the church of Les Invalides, the Pantheon and the Cathedral of Notre Dame itself. The alternative, the alleged intellectual illuministic moderation notwithstanding, was the destruction of the churches. It was during this period that Bishop Henry Gregoire, protesting against the brutality of the revolutionaries, coined the expression "vandalism".

Far from abolishing any kind of spiritualism[4] however, the intention of the Jacobin ideologues with this bizarre "secular cult" of the Supreme Being was to establish a new era in the history of Western religiosity. This was a typically conceited attitude of the Age of Enlightenment which was a complete antithesis of all that had come before. The result was paradoxical, as well as being fully in line with the Chateaubriand's aphorism quoted at the beginning of this section. In claiming the definitive triumph of human reason over religious superstitions, the deists actually ended up transforming rationality into a divine fetish.

However, the Supreme Being, which was totally lacking in history and identity, was immediately perceived as an artificial religion, and had a short existence even within revolutionary France itself. As a phenomenon associated with the Reign of Terror, it followed the shifting fortunes of Robespierre until his final collapse.

This failure did pave the way for further attempts in the same direction, however, the best known of which is certainly the cult of the Great Being proposed by the father of nineteenth-century positivism, Auguste Comte (1798–1857).

---

[3] The cult of the Great Architect of the Universe of Freemasonry, as defined in the Constitutions of Anderson of 1723, is also deistic. Masonic spirituality would require a long treatment in its own right, but undoubtedly represents, at least in some circumstances, a form of secular esotericism, which may be associated with the subjects discussed in this chapter.

[4] In November 1793, during the celebration of the Goddess of Reason, Robespierre burnt an atheistic statue considered "contrary to civic values".

11.4 The shadows of the Enlightenment

As noted by Raquel Capurro, this ideology in fact represented a mere radicalized continuation of the Enlightenment legacy [4]. The two fundamental cornerstones of this theme resumed, i.e., the exaltation of rationalist progress and the cult of an impersonal divine entity, and brought the ideology of the Enlightenment to its extreme, adapting it to the cultural fashion of romanticism.

In the scientific field, this cultural phenomenon led to the birth of so-called "*scientism*", a term with which the same intellectuals of positivistic France indicated their attitude towards the divinization of science. According to Hippolyte Taine, Ernest Renan and Felix Le Dantec, this was a natural outcome. If human reason was the only absolute truth, it had to be independent from any other spirituality whatsoever, as it would be able to provide all the answers humankind needed by itself. Apparently, this was the final affirmation of the millennial Western rationalism, whereas in fact, scientism is the most insidious of irrationalist drifts, being the only one capable of undermining science from within.

Separated from its Christian background, scientific rationalism soon becomes an absolute dogma, which accepts no authority other than its own. From a political point of view, as Karl Popper observed, this path leads to a totalitarian technocracy, which eliminates democracy in order to pursue the good of the community and plans every evolution with engineering precision. From a purely scientific point of view, the greatest risk is the fossilization of paradigms rather than their dogmatization. In celebrating the goals achieved by human reason, science is much less inclined to recognize the limitations, inconsistencies and errors of existing theories, with a self-criticism that, historically, has triggered the great scientific revolutions.

With scientism, the dogma of progress is particularly widespread, an attitude that is unable to recognize the limits of a system or, occasionally, the need to take a step back and then re-evaluate a new direction.

Parallel to these deistic experiments, the Enlightenment and positivism also supported radical atheism, which had among its main exponents Claude-Adrien Helvétius (1715–1771), Paul Heinrich Dietrich D'Holbach (1723–1789), and above all the physician and philosopher Julien Offray de La Mettrie (1709–1751).

The latter, in particular, caused a huge sensation at European level with two essays published between 1747 and 1748, in which he provocatively proposed the concept of "man-machine" and then that of "man-plant", dismantling every spiritual aspect of the human being. The individual was considered to be nothing more than a simple mechanical system, which, while rich and complex, served no purpose other than its mere self-supply. As already noted when dealing with the dark side of imperialism, this de-sacralization of humans led to heavy consequences from a political, social and economic point of view, such as the rise of totalitarianism and the consumerist massification of society. In particular, as Giancarlo Elia Valori observes, *"the man-machine is, as well as a naively scientist hypothesis, a desire for domination over what Nietzsche will call the flock"* [3].

Moreover, taking rationalism to the extreme led to a diametrically opposite outcome, triggering the return of mystical-irrational thinking which occurred in the nineteenth century, as mentioned in the previous pages. Establishing a cause-effect relationship between radical atheism and the return of superstition may seem provocative, but on closer inspection the Judeo-Christian religion represented a synthesis between the need for spirituality inherent in every human being and the peaceful acceptance of the scientific-rational order of Greek origin for almost two millennia, sanctioning the primacy of science over magic.

Enlightenment atheism, while claiming to cancel Christianity at a stroke, in fact had two apparently antithetical side consequences:

- it removed the ethical component from the economic-scientific world, legitimizing the desire for Nietzschean domination, the extreme individualism of De Sade and, in the long run, the totalitarian and imperialist degeneration of the twentieth-century;
- it cancelled the rationalist component of the spiritual world, legitimizing the return of magic and esotericism.

Whereas Christianity had represented a bridge between reason and spirituality, Enlightenment atheism, in eliminating this bridge, instead left the cold utilitarian rationality of totalitarianism and the magical superstition of spiritism, theosophy, anthroposophy and mystical occultism stranded on the opposite banks of the river[5].

On the fertile ground provided by the new-found need for transcendent explanations to the problems of everyday life, supernatural magical thinking was reborn, just as it had happened at the dawn of history. In nineteenth century bourgeois France, the *Book of Spirits* (1857) by Allan Kardec became a true best seller. The book, containing over a thousand questions and answers, had the ambition of gathering the teachings of the spirits of the dead, interrogated through a medium. Starting from this volume, a "spiritualist Pentateuch" was born.

The most emblematic case, however, concerns the international success of the philosopher and medium Eléna Petróvna von Hahn, better known as Madame Blavatsky (from her husband's name).

Well known throughout Europe and the United States for her alleged gifts of clairvoyance, telepathy, dematerialization of objects and mental control, the founder of the Theosophical Society used to ascribe her superior status to her

---

[5] These two apparently opposing degenerative directions are more related to each other than it may seem. A clear example of this connection is the strong presence of esoteric and occult beliefs in the establishment of the German Third Reich. On this topic, see for instance R. Alleau, *Hitler et les sociétés secrètes, enquête sur les sources occultes du nazisme*, Cercle du nouveau livre d'histoire, Paris, 1969; and E. Kurlander, *Hitler's Monsters: A Supernatural History of the Third Reich*, Yale University Press, New Haven, 2017.

contact with *Akasha*, a metaphysical force containing the astral transcription of the history of the universe. Every event that occurred over the millennia would have helped to fuel this mystical energy, which in turn exponentially expanded the capabilities of the chosen few able to get in touch with it. In short, it was an immanent force field that recalls the *Logos* of the ancient Greeks, but with strong Eastern influences.

From the simple paranormal experiences of communication with the world of the dead, Madame Blavatsky moved on to increasingly complex structures, first with her *Isis Unveiled* (1877) and later with *Cosmogenesis and anthropogenesis* (1888). In these works, the esoteric teachings of Hermes Trismegistus, Buddhist philosophy, the Platonic myth of Atlantis and spiritualism merged into a single conceptual framework that sought to replace the Christian system with a systematic "esoteric religion". Steiner's previously mentioned Anthroposophy, which had a heavy influence on the world of science and medicine, originated from a branch of her doctrine, which was known as Theosophy.

The influence of Theosophy on science fiction literature is quite interesting, and perhaps also unexpected. The similarity between the 'Force' of *Star Wars* and the *Akasha* of Theosophy is striking.

In short, the radical rationalism of the Age of Enlightenment had a tangible impact on the Western world, achieving above all in the long run results that opposed the ones that this trend had initially set. The so-called 'magic wave' of the nineteenth century, in fact, has continued to weave its way through history to the present, spreading fascination with the occult and magic even among the intellectual elite of the West.

11.5 PHILOSOPHY VS. TECHNOLOGY: THE DECLINE OF THE WEST?

Since the ancient Greek period, the synergy between philosophy and technology has been a distinctive feature of Western rationalism. Even during times when other regions of the world were more advanced in terms of technology, resources, tools and innovations, Europe retained its distinctive ways of thinking and conceiving reality. However, in most cases, archaeologists, historians and economists tend to disregard the history of philosophy when they deal with the material developments and conditions of society, since philosophy appears to have a limited impact on a material level. The contributions to the history of technology and science provided by individuals such as Anaxagoras, Aristotle, Paul of Tarsus, John Philoponus, Augustine and Scotus Eriugena are often neglected because, strictly speaking, they were never employed to design an instrument or invent a machine.

218   The irrationalistic constant

The picture of the development of technology is thus seriously incomplete. Without taking into account the evolution of human thinking, it is impossible to understand why, for example, Europe was able to colonize most of the planet from the end of the fifteenth century, just after the end of the Middle Ages, without resorting to overwhelming military might. Equally, we would be unable to explain adequately why, starting from the seventh century AD, dynamic and innovative realities such as the Near East and Egypt definitively ceased to play a leading role in the history of innovation.

Philosophy and technology have always represented the two sides of the same coin of Western scientific rationality. Since the 1950s, however, something seems to have changed and these two components, which have always been inextricably linked with each other, now appear to be on a collision course.

Keeping in mind the decline of Hellenistic Egypt, which demonstrates how any scientific innovation can regress to the point of falling into oblivion, it now appears feasible to speak of the "decline of the West", even if this is in a diametrically opposed way to the view of the historian Oswald Spengler[6], who coined this expression.

This is not a certain and inexorable perspective, however. As we have seen in the previous chapters, Western civilization is not in a phase of decline and remains projected "beyond the horizon" by a number of technological and scientific innovations that will likely even overcome the natural barriers of our planet in the coming decades.

However, the danger that the irrationalist constant described in the previous sections will suddenly hold a winning hand is becoming more and more serious. For the first time, many prominent figures in the Western philosophical world are among the supporters of this irrational trend.

As already noted, the transformations of human thinking occur much quicker than the material consequences which unavoidably deeply influence science and technology. Just as many historians and economists are unable to see the seed of the scientific revolution that occurred in the following centuries when looking at the Middle Ages, so, today, there is the risk of underestimating the consequences of the current anti-scientific drift within philosophy.

On the back of the conflict between philosophy and the Christian rationality, the twentieth century also saw an explosion in aversion to technology, a technology which is considered unavoidably dangerous and de-humanizing. One of the most influential Western philosophers of the post-war period, Martin Heidegger (1889–1976), was among the leaders of this trend.

---

[6]O. Spengler, *The Decline of the West*, Alfred A. Knopf, New York, 1962. According to Spengler, all civilizations experience a cycle that is at first progressive and then regressive. By comparatively examining the Western world, it would be possible to conclude that Europe and the United States have been in their declining phase for decades.

11.5 Philosophy vs. technology: The decline of the West?

Heidegger did not actually blame technology as such, but strongly disagreed with its modern developments. According to him, technology has progressively changed its essence over the centuries. In the ancient world, technology, which was limited to the exploration of nature, would progressively have allowed humans to unveil that deep reality that is hidden behind every phenomenon; the philosophical Being.

In the modern Western world, on the other hand, technology is now subject to "Enframing" (*Gestall*), a key term used by the German philosopher to describe an "*organizational principle conceived to satisfy the needs and the will of man, and to do so in the most efficient, productive and systematic way*". In losing sight of the ultimate goal of the whole system, in this new reality human beings are reduced to mere "*technicianized work animals*" [5].

This criticism is obviously aimed at industrial capitalism based on Taylorism and to the semi-robotization of production processes, an undoubtedly burdensome and alienating situation, but one which could only be overcome through a further technological revolution and the introduction of robots and fully automatic devices.

In contrast, Heidegger believed that modern technology was dangerous and unnatural. Technology had lost its original function of unveiling the secrets of nature, and ended up with the far more sinister and unsettling purpose of transforming nature into a resource, a potential reserve of energy ready for human use. Woods became timber reserves, mountains became stone quarries, and rivers became sources of hydraulic power to produce electricity [6]. In short, technology violated nature and prevented it from existing as such, reducing it to a measurable, calculable and exploitable element.

It was evidently a nostalgic and paradoxical thinking, which led historian Andrew Feenberg to point out that "*never has such a succession of non-sequiturs played such an important role in the history of philosophy*" [7]. In fact, Heidegger's thesis saw the ancient world irrationally transformed into a utopian and idyllic dimension.

As we have seen in the previous chapters, the construction of devices and artifacts has always been in response to the aim of automating and simplifying human life. The transformation of nature into "a resource" is certainly not something exclusive to the modern era. The Roman Empire was familiar with the intensive exploitation of mines and land, as well as introducing the artificial into the landscape. Generally speaking, humankind's attitude towards nature has never changed, other than, if anything, an increase in exploitation due to a greater efficiency resulting from capitalist economies.

In other words, Heidegger's view was a romantic longing towards an ideal past that has never existed from a historical point of view. This mystical attitude reached its peak when the German philosopher formulated his own personal proposal for a solution to "the technology problem". His way out consisted of a complete change in the role of science and technology. Citing Holderlin, Heidegger argued that *"where the danger is, also grows the saving power"*. Drawing upon a number of risky linguistic associations, the German philosopher concluded that the model to be considered must be that of ancient Greek technology, the reference point of the contemporary world. In ancient Greek, the verb to make, *poieo*, characterized the technological actions of an artisan, but it was also the root of the word 'poetry', while the term *tekné* referred to both technology and art.

His reasoning was not logical-consequential, but reverted to mere etymological associations. From a methodological point of view, Heidegger's philosophical theses were stated without reasoned argument, and at times even seemed to reveal a purely analogical approach. Nevertheless, his analysis was warmly received in the intellectual world of the West. In legitimizing this view, it became the *a priori* opposition to technology and capitalism, a feeling that is still rampant and has now become a pillar of the so-called post-modern thought.

According to Heidegger and Theodor Adorno (1903–1969), technology should revert to its artistic dimension and away from its economic-capitalist dynamics to constitute a form of speculative, scientific and philosophical exploration [8]. In short, by irrationally glorifying the relationship between humankind and nature, Heidegger and his successors considered technology to be acceptable and not dangerous only when it ceased to have an effective application in the contemporary economy. Moreover, such technology should therefore cease to be technology and should instead become a sort of science (or a natural philosophy) with no practical applications.

The Neo-Luddite movement, an anti-technological movement which, like the early nineteenth century Luddites is prone to resorting to violence, makes a direct reference to Heidegger. Julian Young even goes so far as to claim that Heidegger was a Luddite at the beginning of his philosophical career, and that he supported the destruction of modern technology [9].

In his *Notes towards a Neo-Luddite manifesto*, Chellis Glendinning, one of the founders of the movement, identified electromagnetism (among which are explicitly mentioned television and computers), chemical, nuclear and genetic engineering among the technologies to be destroyed as particularly dangerous [10]. Basically, this list included all the technologies developed since the Industrial Revolution, except perhaps for the most primitive steam engines – provided that they are powered by coal, given that more modern fuels are linked with the chemical industry. In general, the movement advocates a strict application of the precautionary principle in its strongest form.

Theodore Kaczynski, the anti-technology terrorist called 'Unabomber' by the media, has been associated with the most violent part of this Neo-Luddite movement (since it is an informal group, it is impossible to assess who actually belongs to it), and is currently serving life imprisonment because he was responsible for

11.5 Philosophy vs. technology: The decline of the West?

the deaths of three people and injured more than 20 others. In *The Coming Revolution*, Kaczynski argues that to get rid of the current techno-industrial system, it will be necessary to:

- reject all modern technology,
- reject civilization itself,
- reject materialism and replace it with a concept of life based on the values of moderation and self-sufficiency,
- love and revere, or even adore, nature
- enhance freedom,
- punish those responsible for the current situation in such a way that the cost of developing technology becomes too high for anyone to try.

Among those who are to be punished, he includes scientists, engineers, managers, politicians and so on [11].

The Neo-Luddite movement is only the most violent wing (although in the Second Luddite Congress manifesto it is claimed that the Neo-Luddites refrain from violent actions) of a much more widespread tendency that includes philosophers, historians of science and technology, educators, ecologists and sometimes even scientists.

In a post-Enlightenment society in which magic and esotericism are returning to make their influence felt and scientific knowledge is increasingly challenged by so many pseudo-sciences and extravagant and viral conspiracy theories – think of the homeopathic and anti-vaccinist movements, but also the conspiracy theories such as those which deny the historical truth of the Moon landings – technological innovation must also come to terms with these new philosophical trends, which cast disquieting shadows over the future of the Western world.

Is it possible that an era of drastic scientific decline is actually looming? The recent episode of the physicist and mathematician Alain Sokal compels us not to discard this hypothesis. Sokal instigated a prank meant to cause intense debate in the universities of the United States. He unveiled the dangerous tendency among humanistic academicians to handle scientific concepts improperly without necessarily having any knowledge of what they were about. In 1996, the author managed to get an article entitled "*Transgressing the Boundaries: Towards a Transformative Hermeneutics of Quantum Gravity*" to be accepted and published without any objection by the prestigious scientific journal *Social Text*. In fact, the article was completely and deliberately meaningless, and simply put together politically correct notions and mathematical principles with Pindaric flights bordering on the absurd. In one passage, for example, the author even stated that the axiom of equality in mathematics could be considered a homonym of the concept of feminist politics. It was pure 'magical thought', deliberately formulated as such. Sokal bet that the article would be accepted and published without objection, because it "sounded good" and was in line with the journal's progressive political ideology. The facts bore this out.

Reflecting later on this grotesque affair, Sokal observed that the greatest risk for contemporary science stems from the epistemological relativization of knowledge. This philosophical process, excluding the existence of an objective truth and questioning even universally accepted findings, risks completely dismantling Western science by undermining its credibility to the foundations **[12]**.

In the age of the Internet, social networks and means of instant mass communication, this risk is perhaps even more imminent and dangerous. Perhaps the West really is in danger of following the same path which led to the decline of ancient Alexandria in Egypt?

References

1. Eliade M., *Occultismo, stregoneria e mode culturali: saggi di religioni comparate*, Lindau, Turin, Italy 2018, pp. 25-26.
2. Kant I., *Answering the Question: What is Enlightenment?* (1783).
3. Valori G.E., *Spiritualismo e illuminismo* in *Spiritualità e illuminismo*, Futura, Perugia, Italy 2018.
4. Capurro R., *Le positivisme est un culte des morts: Auguste Comte*, Epel, Paris 1998.
5. Bennett F., *Artefactualising the sacred: restating the case for Martin Heidegger's 'Hermeneutical' philosophy of technology* in Deane-Drummond C., Bergmann S., and Szersynsky B., (a cura di), "Technofutures, Nature and the Sacred: transdisciplinary perspectives", Ashgate, Farnham, 2015, p. 51.
6. Giusto N., *Heidegger, Gestell e la questione della tecnica* in "Culture Digitali" (online), 2015.
7. Feenberg A., *Heidegger and Marcuse: The Catastrophe and Redemption of History*, Routhledge, London 2005, p. 22.
8. Babich B., *Between Heidegger and Adorno: airplanes, radio and Sloterdijk's terrorism* in "Kronos Philosophical Journal", vol. VI, 2017, pp. 133-158.
9. Young J., *Heidegger's Later Philosophy*, p. 80. Cambridge University Press, 2002.
10. https://theanarchistlibrary.org/library/chellis-glendinning-notes-toward-a-neo-luddite-manifesto.
11. https://theanarchistlibrary.org/library/ted-kaczynski-the-coming-revolution.
12. Sokal A., Bricmont J., *Intellectual Impostures,* Picador, London 1999, introduction.

12

Beyond the horizon

12.1 FUTURE PERSPECTIVES

As mentioned in the previous chapter, it is possible that the world's civilization is at a crossroads. Could the abandonment of its technological roots and its relapse into irrationalism and pseudo-sciences bring a halt to future developments in the West? Or will this crisis prove to be temporary and allow the Fourth Industrial Revolution to continue towards new goals?

Europe is certainly the weak link in the West. Referring to the continent, Joseph Ratzinger (before he became Pope Benedict XVI) noted in his book written with Marcello Pera:

"There is a self-hatred in the West that is strange and that can only be considered as something pathological. The West tries, in a praiseworthy manner, to open itself to full understanding of external values, but no longer loves itself. In its history, it now sees only what is deplorable and destructive, while it is no longer able to perceive what is great and pure. Europe needs a new – certainly critical and humble – acceptance of itself if it really wants to survive ... Multiculturalism, which is continuously and with passion encouraged and favored, is sometimes above all abdication and denial of what is specific, an escape from its own things" [1].

The West, and in particular Europe, seems therefore to be about to lose its roots which, among other things, we have seen to be essential if it is to continue on its path of technological development.

This denial of its history is accompanied by other purely economic and social phenomena, linked to the growing importance given to finance with respect to industrial production, and therefore to the technology that supports it. This can lead to a reduction of the economic base of the society and a contraction of that freedom that had been so important for the birth and development of modern civilization.

But Europe is not the entire West. While it is not entirely immune, America at least seems to be less affected by these self-destructive tendencies. While the crisis that had affected the Third Industrial Revolution had been linked to the outcome of the Vietnam war and had been, at least initially, a mainly American crisis, it seems that today America has moved on from this and shows signs of recovery, at least in some sectors of the society

After the fall of Marxist rule, Russia is rediscovering its Christian roots and is quickly recovering. It may seem strange to include Russia in the West, but we must not forget that its roots are similar to those of the European civilizations and it began its path towards Westernization centuries ago, despite its ups and downs, as the story of Peter the Great shows us.

Notwithstanding that their roots are very different, Japan and then also China have gone through a very strong Westernization and are now obtaining important results in the technological sciences.

We can therefore hope that the current European crisis will not stop technological and scientific progress, and that the Fourth Industrial Revolution will be able to shift its center of gravity towards the other side of the Atlantic and beyond to the Pacific Ocean. This could prove to be the center of a new development, involving two players considered 'Western' by tradition in North America and Australia, and two that have become 'Westernized', in China and Japan.

In the following paragraphs, we will analyze some possible trends that, if technological progress proves able to overcome the present crisis in the West, will lead us to new developments.

12.2 BEYOND THE TECHNOLOGY (AND THE ECONOMY) OF OIL

Since the beginning of the Industrial Revolution, industry has been based around the combustion of fossil fuels to generate the energy required for the production of an ever-increasing number of goods. Even the energy required for non-industrial uses (heating, agriculture, transportation) was, and still is today, largely obtained by burning fossil fuels – first coal and then oil and natural gas.

12.2 Beyond the technology (and the economy) of oil

This has obviously led to the increased use of fossil fuels and, particularly since the Second Industrial Revolution, of oil. We have already addressed the Neo-Malthusian vision, which states that one of the major short- and medium-term problems will be the scarcity – or even the exhaustion – of raw materials in general and of oil in particular. We have also observed that at present, the discovery of new reserves is proceeding at a faster rate than the exhaustion of the old ones, which is why this scenario is highly unlikely. The problem is that the cost of extraction increases as the deposits which are easier to extract are exhausted, and that, at a certain point, it will be necessary to switch to alternative technologies for economic reasons. Saudi Arabia's former oil minister, Ahmed Zaki Yamani, was referring to this situation when he made his famous quote, reported by the Reuters agency in 2000. The full quote was:

"In the next thirty years there will be a large availability of oil and no buyer. The oil will be left underground. The stone age did not end because there was a lack of stones, so the oil age will not end because oil will be exhausted".

The actual availability of oil in the near future is a very controversial issue, and present indications range from the alarming statement that the world's oil reserves are almost exhausted (or at least that the peak of production has already been reached) to the reassuring declaration that there will be no problem with the supply of oil in the twenty-first century. Such statements, and all the possible intermediate nuances, are often presented as factual data, to such an extent that the increase in oil prices in the recent past has been interpreted by some as evidence of the most pessimistic predictions.

An estimate of the energy consumption in the twentieth century and a forecast prepared by the U.S. Department of Energy for the twenty-first century is shown in Figure 12.1. The consumption was reported in billions of barrels of oil equivalent per year, regardless of the actual nature of each of the energy sources considered. Other forecasts report much less pessimistic results for oil, but above all for natural gas which at the moment seems to be much more abundant.

A critical point of this scenario, and of many other similar ones, is the hypothesis that research into the production of nuclear fusion energy will not provide any concrete results in the twenty-first century. The scenario would look very different if nuclear fusion energy could be made available in good quantities within a few decades. In addition to nuclear fusion, none of these scenarios take into account any technological developments that could reduce energy consumption or make new energy sources available. However, while it is not necessarily wise to count on future developments that may not materialize, this does often lead to quite pessimistic scenarios.

226 Beyond the horizon

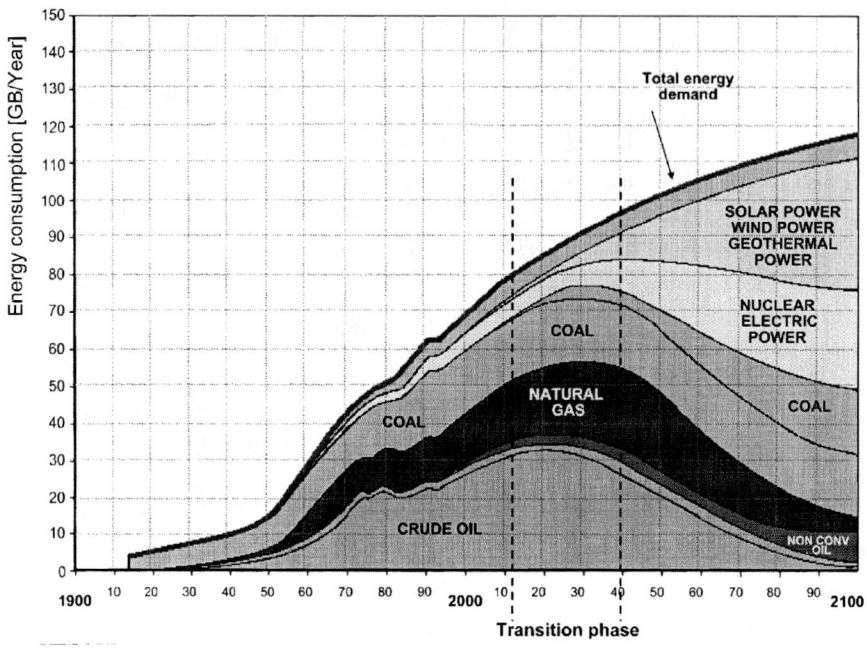

Figure 12.1. Evolution of energy consumption in the 20th century and forecasts for the 21st, according to the United States Department of Energy. [Image from G. Genta, A. Genta, *Motor Vehicle Dynamics, Modeling and Simulation*, World Scientific, Singapore 2017.]

Regardless of which scenario materializes, it will be necessary to develop alternatives to fossil fuels, in particular nuclear energy and solar energy (which includes photovoltaic, solar thermal, wind and hydro power, which ultimately come from the Sun).

However, when we talk about replacing one raw material with another, we should not just examine its costs and availability, but also look at its quality. One such example, often cited, was the replacement of whale oil with lamp oil for home lighting in the nineteenth century, due to the increase in the cost of the former thanks to excessive whaling. Lighting certainly became cheaper, but lamp oil produced more domestic pollution and health problems, as well as much less comfort due to the unpleasant smell.

The production of energy through fossil fuels is creating problems, not so much related to the scarcity of oil as to the large amount of harmful substances released by combustion in the atmosphere. These range from pollutants (nitrogen oxides, carbon monoxide, etc.), essentially due to imperfect combustion, to greenhouse gases, mainly carbon dioxide.

12.2 Beyond the technology (and the economy) of oil

As far as air pollution is concerned, at least in theory, it is possible to achieve completely clean combustion (zero emissions) through improvements in the combustion process and the treatment of fumes, in order to avoid emitting those polluting substances that cannot be eliminated at the source. However, little can be done with regard to greenhouse gases, because for any given fuel – with the exception of pure hydrogen – the carbon dioxide emitted is proportional to the energy produced by combustion. The only way to reduce the greenhouse effect is to increase the efficiency of the energy production process, and thereby reduce fuel consumption.

Essentially, all fuels contain carbon and hydrogen atoms – with the exception of coal that contains only carbon – and therefore perfect combustion necessarily produces carbon dioxide and water. In addition, if they contain other substances that we define as harmful, such as sulfur, they produce pollutants such as sulfur dioxide. For example, burning coal produces 0.37 kg of carbon dioxide per kWh produced. The ratio is 0.26 for gasoline, 0.25 for diesel fuel and 0.18 for methane, excluding any polluting substances due to impurities in the fuel and poor combustion.

To reduce the emission of greenhouse gases drastically, a first solution is to cease using processes based on combustion to produce energy and replace them with sources such as nuclear energy, as well as those normally referred to as renewable energies which come substantially from solar energy, either directly, as in the case of photovoltaic panels, or indirectly, as in the case of wind, hydroelectric energy, etc[1].

Things are complicated by the fact that many of the latter require a considerable energy expenditure for the construction of the power plant. This means that when the greenhouse gas emissions of the various technologies are compared, we must take into account not only the carbon dioxide produced during the use of the plant, but also of that produced to build it.

Another option is to produce fuels through biological processes (biofuels). The production of alcohol from the fermentation of agricultural residues, for instance, emits exactly the same amount of carbon dioxide during combustion that the plants from which the biofuel has been produced had taken from the atmosphere during their life. It is said, therefore, that biofuels do not produce greenhouse gases.

In the future, biofuels will probably be produced mostly by genetically modified bacteria in large bioreactors using processes similar to photosynthesis (artificial photosynthesis) which, starting from the sun's energy, water, atmospheric carbon dioxide and agricultural waste of various kinds, will produce methane and then more complex hydrocarbons. In any case, a mix of various technologies will be required to replace the oil economy, including nuclear energy, the so-called renewable sources and biofuels. If controlled nuclear fusion becomes possible, it

---

[1] Another important source of renewable energy is geothermal energy, but in this case, the energy comes mainly from the radioactive decay of some substances that have been part of the Earth since its formation.

will permit the production of large quantities of energy at low cost. The percentages of this technology mix will be subject to local factors and the continuous technological developments on which the economic benefits of the various sources will depend [2].

Oil will remain an important source of raw materials with which to create a wide range of chemical and pharmaceutical products. In fact, oil is too precious as a raw material simply to be burnt to produce energy. This will remain the case at least until the same chemicals can be obtained more cheaply from genetically modified bacteria in bioreactors. Then, Yamani's predictions will be realized, and the oil will remain underground.

12.3 ELECTRONICS AND TELECOMMUNICATIONS

As mentioned in Chapter 10, the essential elements of the Third and Fourth Industrial Revolutions were the development of electronics and telecommunications, the technologies that are normally called ICT[2].

The origins of electronics can be traced back to 1906, when Lee De Forest invented the thermionic valve and in particular the triode. Thermionic valves enabled the development of a large number of simple electronic systems, from the radio to the first computers. However, it was only with the introduction of solid state devices, and in particular of transistors which entered the market in the 1950s, that miniaturization began, leading to the possibility of creating increasingly powerful but less bulky circuits. With microelectronics, it became possible to provide the machines with electronic control systems and an actual artificial intelligence.

Since 1906, the size of electronic systems has decreased by a factor of ten million, and between 1946, when the first electronic computer was built, and the present-day microprocessors of modern smart phones, the performance per unit of required power has increased by a factor of 10^{12}, i.e. one thousand billion times[3]. For example, a 1 Tb (terabit) memory can be purchased for $100–200 and can fit inside a mobile phone. Using the technology of 110 years ago, the equivalent capability would have a cost roughly 100 times the gross national product of Japan, weighed 100 million tons (equivalent to something like 20 million elephants), and

---

[2] This section is inspired by Hiroshi Iway's lecture *End of Miniaturization in Electron Devices and World after That*, presented at the International Conference on Materials and Computers in Science and Industry 2017, Corfu, August 2017.

[3] We must be extremely careful with very large numbers, because Europe and the USA use different scales (long scale for the former, short scale for the latter). For example, using the short scale (USA), one trillion is 1,000,000,000,000 or one million, million, or 10^{12}, while using the long scale, it is 1,000,000,000,000,000,000 or one million, million, million, or 10^{18}.

required a power of 50 billion kW (or about 50,000 large nuclear plants) to power it. Basically, the computer revolution that has given us computers, cellphones, the internet and all the things we take for granted today would not have been possible without the miniaturization of electronic devices.

The tendency towards miniaturization, and therefore to the increase in the power of electronic devices with the same volume and weight, was formalized in 1965 by the so-called Moore's Law[4], which stated that the complexity of a microcircuit, measured for example by the number of transistors per chip, doubles every 24 months. In fact, Gordon Moore, co-founder of Intel, initially talked about doubling every 12 months, but regardless, what matters is that the complexity (or power) of microprocessors grows exponentially over time (Fig. 12.2).

Despite the term, Moore's law is not a physical law, but an empirical observation of an industrial phenomenon, one that has, at the same time, 'physical' and economic components, so it cannot be taken for granted that this growth will continue in the future.

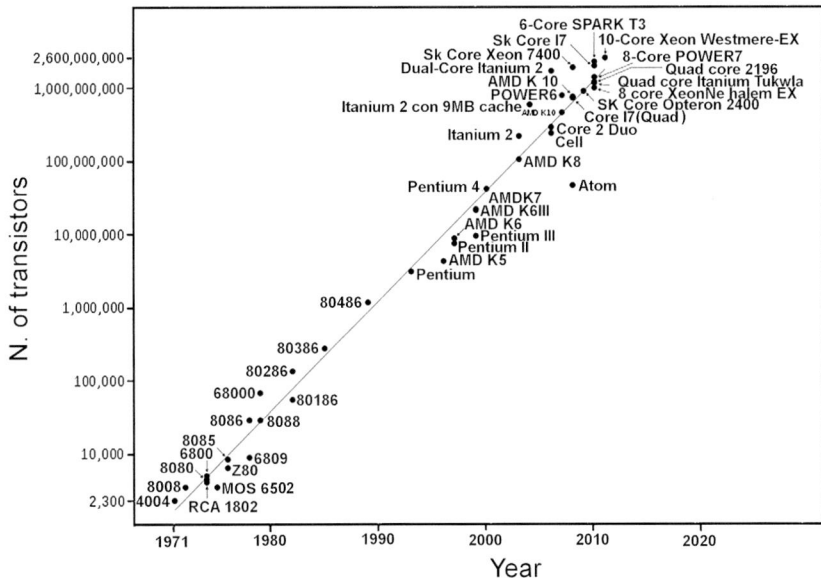

Figure 12.2. Moore's law, showing the increase in the number of transistors in a microprocessor as a function of time. Note the logarithmic scale. [Image modified from upload.wikimedia.org/wikipedia/commons/2/25/Transistor_Count_and_Moore%27s_Law_-_2008_1024.png.]

---

[4] A second Moore's law deals with the increase in the cost of the plants needed to produce microprocessors.

Obviously, no material device can have a size smaller than that of an atom – we speak of a size of the order of one nm (one millionth of one mm) – so there are well-defined limitations to the miniaturization of electronic devices. It is thought that this limit will be reached in the short term, probably by 2025.

But this does not mean that electronic devices must necessarily stop progressing or that the electronics industry in the future would enter into a stagnation that would prevent it from fulfilling those promises that have been discussed for years. Certainly, we can expect progress in the field of smart phones and other portable devices, as well as in robotics, both industrial and service, even if the predictions that the robotics industry in the twenty-first century would play the propulsive role that was typical of the automotive industry in the twentieth have yet to materialize. That same automotive industry will see important applications of electronics, in vehicle electrification, for which power electronics will play a key role, and in driving assistance and then autonomous driving. For the latter, artificial intelligence will play a significant role, as it will in the management of an increasing amount of information (the so-called 'Big Data') and in the management of energy, mobility, and other fields (smart grids, smart cities ...).

There is a great debate between the supporters of strong artificial intelligence and those who do not think it is viable. We must underline that the basic dogma of strong artificial intelligence – the affirmation that thought is an algorithmic process and therefore it is possible to realize truly thinking machines and even self-conscious machines – remains precisely that, a dogma which has never been proven. There are those who think that artificial intelligence has nothing to do with human intelligence and that machines, at least those achievable with predictable technologies, will never think or become self-conscious.

Many argue that advances in the field of computers will soon lead to a singularity. In this context, 'singularity' means a condition in which technological progress accelerates to the point that it is no longer possible for human beings to understand and follow further developments. According to Ray Kurzweil and the transhumanists, this will occur because artificial intelligence will, at some point, become greater than human intelligence and machines will become the "*dominating species*" of this planet [3]. The only way left to humans would be to incorporate these thinking machines into their own bodies, giving way to hybrid man-machine beings, the so-called cyborgs.

The limitations to Moore's law mean that we cannot continue with an exponential growth in the power of computing machines and consequently that even the so-called singularity is nothing but a myth. In order to continue the progress in electronics after reaching the limits of miniaturization, there are two potential paths: connecting separate machines into a network and distributing computing power between various units (cloud, as an Internet extension, and Internet of Things, IoT); or changing the technology or even the very way of working for computer systems (quantum computers). However, even if these options make it possible to continue beyond the limits inherent to Moore's law, these new technologies themselves will also have limitations, so that it would not be possible to continue with an exponential expansion. Even in this way, we cannot reach a 'singularity'.

Another possible change in technology would be to shift the evolutionary line from physics to biology. Consider that the microscopic brain of a mosquito, consisting of a few neurons concentrated in a very small space, is able to process images in real time while at the same time controlling the flight of the insect using a very small amount of energy. Biological computers are substantially more efficient than those based on semiconductors. One can thus think of using software based on the animal brain on more or less standard microprocessors, or of introducing the brain, or parts of the brain, of insects or other animals into computational machines, which would become partially organic. Moreover, the animal brain assembles itself by means of the instructions contained in the DNA and can be modified by alterations to the latter. In the long term, we can consider making microprocessors with semiconductors talk with biological computers and thus exploit the advantages of the two ways of processing information, but this will probably take a very long time (Fig. 12.3).

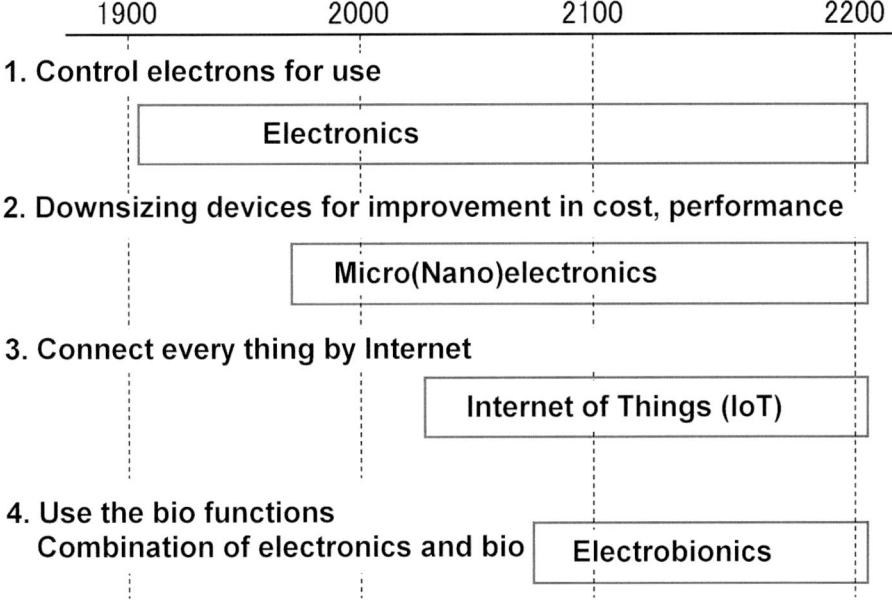

Figure 12.3. From electronics to electrobionics.

12.4 BEYOND EARTH

As mentioned while dealing with Neo-Malthusianism, we are still very far from the hypothetical situation of a scarcity of raw materials, and the recurrent price increases of some goods are linked more to political and speculative phenomena than to real crises due to their scarcity. However, the situation is not as bright as it might seem:

- Because the more easily accessible deposits are depleting, it is necessary to resort to deposits that require higher extraction costs or to less-rich minerals, which means a larger quantity of ore must be refined to obtain the same quantity of product.
- Pollution produced by mining operations increases as more difficult deposits are used, which further increases the cost of extraction and refining, because at the same time legislation becomes more restrictive in terms of environmental protection. A very important example is neodymium, a crucial component of magnets for electric motors, which is becoming increasingly essential as automotive electrification progresses. There are large neodymium deposits in Canada and the United States, but in those countries the cost of production is very high due to anti-pollution legislation, to the point that it is impossible to extract it conveniently. The only sources of neodymium currently usable are in China, thanks to a less restrictive legislation, which is why Chinese producers are currently operating as monopolies. However, sooner or later the legislation of that country will impose more restrictive standards.
- Finally, even if it is in a distant future, the natural resources of our planet will eventually run out.

The raw materials which we need for our technological development are abundantly present on the celestial bodies closest to us, such as the Moon and the so-called NEAs (Near-Earth Asteroids), where environmental conditions are so extreme that any human intervention could hardly make things worse. On the contrary, by extracting material from the NEAs, the danger that these bodies present for the Earth and all the forms of life that exist on its surface would be reduced (the extinction of dinosaurs, caused by the fall of an asteroid, reminds us that this is a danger with low probability, but it is always present).

A single metallic asteroid of medium to small size, for instance, contains quantities of metals of the platinum group (gold, platinum, iridium) and the iron group (iron, cobalt, nickel) comparable with the entire terrestrial production of several decades. While the latter are metals of technological interest, the former are precious metals and one objection to extracting them from extraterrestrial sources is that their value is intrinsically linked to their rarity. Having them at our disposal in large quantities would drastically reduce their price, creating problems not unlike those experienced in Spain when the import of precious metals from the New World caused uncontrollable inflation.

In fact, the fall of the price of precious metals in this case would have extremely beneficial effects, because it would transform them from precious metals into technological metals, allowing them to be used in many applications. Just think of the importance of platinum as a catalyst, for example in the exhaust systems of cars, or the possibility of using gold in electrical connectors where its properties as an excellent conductor would be very useful, and which is not possible at present because of its cost.

The use of extraterrestrial resources is not just about metals, precious or otherwise, but also concerns a great quantity of other resources. These include water, which is abundant on comets and present on both the Moon and Mars – water is precious in itself, but also for the production of propellant – rare earths, which are indispensable for electrical applications (as mentioned, the cost of neodymium is currently one of the factors that hampers electrification in the automotive field), or the He3 isotope of helium, present in small quantities on the Moon (but completely absent on Earth) that could become a key asset for the production of energy in nuclear fusion power plants. Given the enormous amount of energy that can be obtained from very small quantities of helium 3, the cost of its transportation from the Moon would already be marginal now, and once we are be able to operate nuclear fusion power plants, it would completely replace fossil fuels.

However, the space economy is not based only on raw materials which could be obtained in abundance and at decreasing costs in the near future [4]. Space today is already economically very important for supplying services, from telecommunications to the management of terrestrial resources, from meteorology to the management of emergencies. Earth's economy is now so dependent on the services provided by satellites that it can no longer do without them [5].

Another field is that of energy production. Satellites equipped with large photovoltaic panels can produce energy from the Sun in large quantities, energy that can easily be conveyed to Earth by microwave beams. This is a completely non-polluting energy, even from the point of view of global warming, as well as being low cost once the power plant is built. It is an energy that can be added to that produced by other renewable sources and nuclear power plants, for an economy completely free from any dependence on fossil fuels [6]. Currently, what is hampering this technology is the cost of launching the satellites needed to produce the energy, but these costs could be reduced thanks to progress in the field of launch vehicles – the privatization of launch services has already led to large cost reductions in the recent past and this effect is bound to continue – and above all with the possibility of using materials obtained from the Moon for their construction, which would easily be transportable to Earth orbit due to the low gravity of our satellite.

Another key point on which space economy is based, and which is likely to become its main business model over the coming years, is space tourism. There are many people, rich enough to pay for flight in space, who are extremely interested in experiencing the thrill of a visit to a space station and, above all, in visiting the Moon and eventually Mars.

Of course, this is a tiny percentage of the inhabitants of Earth, but in absolute terms these are people who, thanks to their enormous financial resources, would represent a large part of the space economy. The phenomenon is not new. As noted previously, thanks to their enthusiasm for mechanics and watchmaking, a few members of the French and especially the Russian nobility provided a market in the seventeenth and eighteenth centuries that allowed the development of innovative machine tools. Such machines could not have been developed solely for the artisans, given their modest financial resources, but once built, they enabled great progress in technology and industry. Space tourists are now able to finance similar innovations in launching, life-support and mobility systems that will allow space exploration and the space economy to develop much faster than is possible by focusing on scientific research alone and on the needs of the inefficient public organizations, the space agencies.

Clearly, human expansion in space will affect the nearest celestial bodies in the near future, such as the Moon, Mars and some asteroids, and will occur in rather hostile and difficult environments. As technology progresses, it will become possible both to explore and colonize more distant celestial bodies, and to modify the environment of many nearby celestial bodies to make them truly habitable. This process, known as terraforming, would be applicable to a planet like Mars, where it would arouse some ethical concerns given that any indigenous life forms (at the level of bacteria or other very simple forms) could be damaged and even brought to extinction by these operations, but would be more difficult for the Moon where there are no such ethical problems.

If humankind learns to move from one star system to another before succeeding in terraforming any planet of the solar system, it is possible that we will instead look for more distant habitable planets. Otherwise, the solar system will be populated first, possibly creating mining colonies on the asteroids and on some satellites of the outer planets, and then we will expand beyond. The result, whichever way round it is done, will be to create a multi-planetary civilization.

The rocky planets, the rocky satellites of the planets, or the asteroids of the solar system have their lithosphere (by definition), are often devoid of atmosphere or have a very faint atmosphere, but in general do not have a hydrosphere (except for Titan, a satellite of Saturn, which has a hydrosphere made of liquid methane) and, as far as we know, do not have a biosphere. As soon as colonization begins, they will begin to have a biosphere (in protected environments), a noosphere and a technosphere. Strictly speaking, the Moon and Mars already have a technosphere, consisting of the few artifacts that have been left by the various space missions.

To transform a planet, therefore, will mean modifying its hydrosphere and atmosphere (or creating them if they do not exist) so that they can host a terrestrial biosphere.

12.5 THE SECOND NEOLITHIC REVOLUTION

As a result of the Neolithic Revolution, humankind learned to produce its food by cultivating plants and raising animals, instead of collecting and hunting the living beings that developed spontaneously in the environment in which they lived. The Neolithic Revolution deeply changed not only human life, but above all the planet Earth. While it enabled an incredible increase in the number of people who can live on our planet, it also caused an even greater increase in the cultivated area. The process of photosynthesis which, thanks to sunlight, transforms inorganic material (water, carbon dioxide and other substances) into food (organic material that forms the body of living beings and is the fuel that supplies them with energy) has an extremely low efficiency: no more than 1–1.5 percent. If, thanks to breeding, we want to obtain animal food, then the yield of the chain from sunlight to food ready to be eaten is even lower. Actually, only a small fraction of the sunlight reaching our planet is converted into 'ready to eat' chemical energy.

Hydroponic crops, and even cultivations in which the light required for the development of plants is provided by LEDs (Light Emitting Diodes) instead of directly by the Sun, are now a reality. It should be noted, for these purposes, that the introduction of LED lamps and high-efficiency solar panels is important. By turning sunlight into electrical energy and then the latter into light of the wavelength to which the plants are more sensitive — and thanks to the fact that LED lamps produce very little heat and can therefore be placed directly on the leaves — the total yield is much higher than in the case of conventional crops. This means that, for the same agricultural production, the area required by solar panels is much smaller than that necessary for conventional crops. With further measures, such as enriching the atmosphere with carbon dioxide and particular nutrients, the yield of the crops can definitely be increased.

But this is only a first step. Studies have been underway for a number of years to produce wholly synthetic food from inorganic material through the use of electricity. These are research projects generally funded by foundations, such as the Bill Gates Foundation and similar philanthropists, due to the importance of these researches in reducing the dangers of famine and other problems related to possible food crises.

The basic idea is to give the food thus obtained the same appearance as we are used to, but it is also possible to improve its characteristics. For example, synthetic meat could be designed to have a lower cholesterol content than its natural equivalent. However, these are organic substances produced with biological processes, starting from genetically modified micro-organisms or from stem cells of animal origin.

Once the basic substances have been produced, and possibly adapted to the needs and nutritional and medical problems of the individual customer, three-dimensional printers would be able to prepare the food suited to individual tastes.

Although the production of synthetic foods gets announced from time to time, for now it is carried out at a research and experimentation level and will still take a relatively long time to reach mass production, particularly because the cost of these synthetic foods is higher than that of conventional food for now. It is probable that initial applications will be linked to expansion in space. Even today, producing artificial food on Mars from local ingredients would cost less than bringing ready-to-eat food from Earth. Over a timescale that is not yet easy to predict, these technologies will spread, causing a revolution on a scale similar to that which took place about ten millennia ago which we call the Neolithic Revolution, to the point that we could call it a 'second Neolithic Revolution'.

Or perhaps it would be better to call it a 'Neolithic counter-revolution', in the sense that over time, our planet could return to being the 'wild' and 'natural' place it was until the Mesolithic. Cultivated fields could disappear, giving way to forests and grasslands, while the domesticated animal species could give way to the wild animals that preceded them.

12.6 MANY HUMAN SPECIES

In the middle, or at worst, the end of the twenty-first century, a small but significant portion of humankind will live outside planet Earth, in space stations, on the surface of the Moon and Mars, or perhaps even on celestial bodies even further away, such as the asteroids or the satellites of the outer planets. In all these 'places', the conditions will be quite different from those on Earth, particularly in terms of gravity, whether much lower than on Earth or even almost none. While it would be possible to generate an artificial gravity in the case of colonies in space through rotation of the habitat, this would be much more difficult in settlements on a celestial body of any kind. On the other hand, even initially, there will be a considerable advantage in maintaining a gravity lower than that of the Earth.

On Earth, the most isolated communities, particularly those which have remained so for thousands or tens of thousands of years, have not physically evolved to give rise to a new species. In fact, the somatic differences that have occurred over the last millennia are so small that anthropologist Stephan Jay Gould has estimated that a Cro-Magnon man brought into our society would not show differences in behavior or attitudes other than those due to cultural evolution. In expanding beyond Earth, however, something very different may happen. Differences in the natural environment could lead humans to diversify more and more until they give rise to different species. The fact that relatively small communities would be isolated in different environments is certainly a factor that would increase the probability of mutations, as would living in an environment where the radiation is more intense than on Earth.

Such diversity would not be limited to humans living on the planets and humans living in space habitats, but also to different human species living on planets in different star systems. As suggested by the ethnologist Ben Finney, humankind could differentiate itself sufficiently to produce actual aliens, of human descent [7].

However, it can be assumed that this 'Darwinian' evolution of the human species, which would require a timespan of geological proportions, is unlikely. Evolution proceeds by mutations and natural selection, and humans, ever since they have had the opportunity, have always tried to overthrow natural selection. After all, the main task of medicine is precisely that: avoiding the application of natural selection to our species, to guarantee the greatest chance of survival for all individuals regardless of how favorable their genome is. *Eugenic* practices, which would facilitate selection among the individuals of our species, are ethically rejected by almost everybody and are certainly unlikely to be accepted in the near future.

However, even if it is unthinkable that the evolution of the human species would start again with expansion into space, creating different species adapted to the various planets, and even though humankind would instead do everything to adapt the environment to its current form, there will be strong pressures to change the genetic heritage of individuals in this sense.

As soon as there are permanent communities living in conditions different from those usual for humans on Earth, a very serious problem will arise: To what extent is it acceptable to intervene in the human genome to adapt people artificially to the environment? Certainly, it would be much simpler and cheaper to adapt the inhabitants of a space station to a lower pressure than to build suitable structures to maintain the usual pressure, and this is just the most trivial example. There is no doubt that such adaptation would be possible in the near future, but to what extent is it permissible to go along this path? It should be noted that such decisions might also be taken for reasons that are political rather than strictly technical. Deciding to adapt the inhabitants of a number of space stations to new conditions may mean preventing new inhabitants from arriving on them, or at least discouraging them from doing so. Differentiating the species could also be conceived as a means to obtain political ends, masking them with a rationalization aimed at improving the conditions of life.

Up to now, this aspect of bioethics has not aroused great interest, since it is intrinsically linked to expansion in space, but it is important that our civilization is culturally prepared to face the moment when certain perspectives open.

It must also be noted that any push in this direction would depend on technological perspectives. If interstellar travel of any kind remains only a distant future possibility for centuries, then colonization will focus on the very hostile environments of the solar system, such as space itself or asteroids, and pressure to adapt humans will be maximum.

On the other hand, if the human species learns to move from one star system to another with extremely long colonization journeys – here, we are talking about space arks or generation ships, which are large spacecraft in which generations and generations of people would spend their whole lives on a long journey lasting hundreds of years to move from one system to another – then terrestrial planets could be opened to colonization and things could become easier from the point of view of the environment in which the settlers will have to live.

However, contact between the various systems would likely be very scarce, and so possible differentiations between the various human groups could have limited consequences. All in all, does it really matter if the inhabitants of a system that is astronomically close, but very different according to human parameters, end up being quite different from us and actually belong to a different species? In the anthropological museums, images of different human species may well accumulate gradually, a subject of study for the specialists and a curiosity for those interested in anthropology, but a fact of little relevance in everyday life.

Even before these possibilities are realized, things may be different if it is discovered, as some have already claimed, that it is possible to travel at speeds higher than the speed of light[5] and that terrestrial planets could be reached relatively quickly, even if they are very far from Earth. In this case, the incentives to differentiate humans could be smaller, but easier contact between the various star systems could amplify the importance of these differences.

References

1. Ratzinger J., Pera M., *Senza radici. Europa, relativismo, Cristianesimo, Islam*, Mondadori, Milan, Italy 2004.
2. Saracco G., *Chimica Verde 2.0*, Zanichelli, Bologna, Italy 2017
3. Kurzweil R., *The Singularity is Near*, Viking press, 2005.
4. Sommariva A., Bignami G., *L'economia dello spazio*, Castelvecchi, Rome, Italy 2017; Bignami G., *Oro dagli asteroidi e asparagi da Marte*, Mondadori, Milan, Italy 2015.
5. Johnson L., *Sky Alert!*, Springer, Chichester 2013.
6. Mankins J.C., (Editor), *Space solar power*, International Academy of Astronautic, Paris 2011.
7. Salomon W., *The Economics of Interstellar Commerce*, in *Islands in the Sky*, Wiley, New York, 1996; Finney B., *From Sea to Space*, The Macmillan Brown Lectures, Massey University, Hawaii Maritime Center, 1992; Finney B., *The Prince and the Eunuch*, in *Interstellar Migration and the Human experience*, University of California Press, Berkeley, 1985.

---

[5] Actually, relativity theory shows that it is not possible to move at speeds greater than that of light. However, some solutions, derived from this theory, would enable the exploitation of some peculiarities of space-time to bend it, reducing the time for an interstellar journey to something less than the time taken by light for the same journey.

Epilogue

We cannot deny that the human species has come a long way in the nearly two million years it has been on this planet, nor that what has allowed human beings to transform themselves *"from animals to gods"*, as Yuval Noah Harari would say, has been technology.

We can identify two main turning points in this journey. The first, which occurred about 50,000 years ago, was the cognitive revolution, when humans took the first step towards abstract thinking. Perhaps because of an improvement in their hardware (the brain, the larynx), perhaps due to an improvement in their software (the ability to think, the refinement of language), but probably due to a simultaneous improvement of both, they learned to give a name to things that could not be perceived through their senses and of whose very existence they could not even be sure, and to reason on them as if they were real material objects.

It should be noted that the cognitive revolution has little to do with technology. Human beings chipped stones, controlled and lit fire and shot arrows well before reasoning on their gods or making use of a number of absolutely trivial things, which they used as standards to measure wealth, to exchange for more useful objects – another abstract concept. Perhaps money itself was the only technology that came directly from the cognitive revolution.

The second stage, this time purely cultural, occurred much more recently at about a couple of millennia ago, and took place in a very gradual and almost imperceptible way. Humans invented a method to think rationally about the reality that surrounded them and later learned to apply this method to technology. That very technology, which allowed them to build the tools that enabled them to compensate for the deficits of their physical structure, produced an evolution that has led to a 'generalist' animal, without any particular specialization.

One of the effects of science has been the continuous accumulation of knowledge. Before science, technology was based on a process of trial and error, which was poorly documented or even impossible to document, because the innovators themselves did not know why or even how the innovative processes worked. The progress of technology was a slow succession of steps forward and steps backwards.

Later, with the development of scientific method, each generation started from the certainties discovered by the previous ones and built upon them. Even when a theory was demonstrated that replaced an earlier one, the old theory was not forgotten but became a special case or an approximation of the new, with its field of validity circumscribed but not canceled.

This does not mean that the details of a technology that is unused for any length of time cannot be forgotten, nor that it would not require considerable effort if they are to be reinvented. However, even in this case, humans no longer fall back into the myth of forgotten technologies, of a lost wisdom of mythical 'ancients' as was common in the past. The increasing role of science in technology therefore caused a noticeable acceleration in the pace of progress.

At the conclusion of this discussion on the importance of technology in the development of humankind, an obvious question comes to mind: Was it worth proceeding along this path? Or, to use Harari's metaphor, was it worth turning ourselves from animals to gods? Perhaps this question does not even make sense, because no one has ever chosen to go along this path, and no one has ever asked us whether we wanted to remain animals or wanted to become gods. For most of this process, indeed, no one even noticed that this transformation was taking place. Up to the end of the classical world, the general perception was that no change was going on, or that if there was a change, it was for the worse or it was a cyclic change leading nowhere.

If we consider that the behavior of each individual, belonging to any species, is determined by the tendency of the species to spread, occupying as much as possible of its ecological niche and possibly also expanding outside it, then the human species has had an amazing success, showing that the strategy of 'generalist animal' is a winning one. Perhaps too much so, judging from the number of species whose extinction we have caused. But this is precisely the goal of the egoism of genes (in the figurative sense introduced by Richard Dawkins in his famous essay [1]). From this point of view, one might think that intelligence, and then technology, are nothing more than tools of our genes which will enable us to expand even beyond the barriers of our planet, potentially colonizing an endless number of planets, something that no other living being has ever managed to do as far as we know.

Obviously, this way of seeing ourselves and our role in the world does not satisfy us at all. The idea of merely being the instruments of the will and of the lust for power of our genes – in a figurative sense of course, because genes cannot have either a will or a lust for power – does not correspond at all to the idea we have of ourselves and it is, so to speak, humiliating.

Let us try to make a balanced assessment of our evolution, in particular from the Mesolithic onwards.

From a material point of view, each of us, at least in developed and emerging countries, has at his disposal a greater quantity of material goods available than probably all but the very highest echelons of society over the last few centuries. Life expectancy at birth has also greatly increased, although here we must strike a note of caution. It has often been said that, in 'natural conditions' the average life of human beings was very short, and this is probably true. This is mainly linked to the effects of very high infant mortality rates on the average lifespan. Any of our hunter-gatherer ancestors who were lucky enough to survive the first years of life had a fair, if not quantifiable, probability of reaching a respectable age. The remains of men and women who had died in their fifties or sixties, as well as some older still (perhaps very fortunate cases), have been found. Things probably got worse with the Mesolithic Crisis, and even after the Neolithic Revolution average life span did not increase. Indeed, in many periods it seems to have diminished. The growing dependence on agriculture, and therefore the susceptibility to natural events such as droughts, floods, and harvest fires, or to catastrophes produced by humans themselves such as raids, wars, and invasions, made human life precarious in many periods. Later, the overcrowding of cities definitely worsened the health situation, at least for the lower social classes. The result of all these circumstances seems to have been decreases to the average life span. For example, while values of 33 years are often mentioned for the Paleolithic era, lower values are mentioned for later periods: from 20 to 33 for the Neolithic; 26 for the Bronze Age; from 25 to 28 for classical Greece; and from 20 to 30 for the Roman period. These values seems to have lasted until the modern era, when scientific medicine increased life expectancy again, in part thanks to the enormous reduction in infant mortality and mortality at childbirth, and also in terms of the probability of life for those who passed puberty. The graph in Figure 13.1 refers to Great Britain, which can be considered representative of the industrial countries.

As far as the rest of the world is concerned, the values are generally lower (though not everywhere), but they are also increasing. The overall world average for 2015 was 71.86 years, compared to 81.61 for Britain as shown in Figure 13.1.

Perhaps the circumstances relating to the decrease in life expectancy from the Paleolithic to the Neolithic are influenced by the myth of the healthy and happy life of our hunter-gatherer ancestors – a sort of modern version of the 'Golden Age' myth – but they certainly have a foundation.

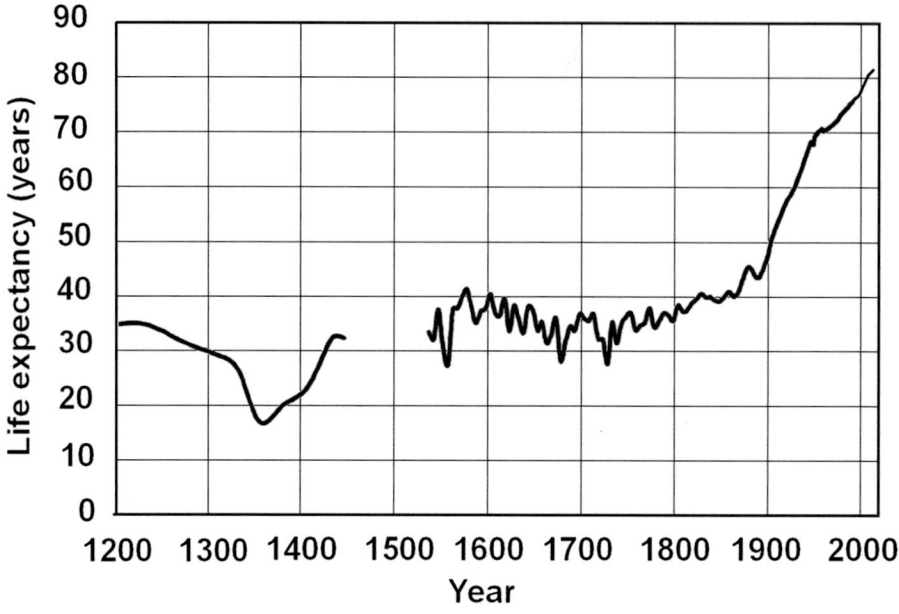

Figure 13.1. Life expectancy at birth. 1200–1450: England, male landowners (Bjørn Lomborg, *The skeptical environmentalist*) [2]; 1541–1998: Great Britain, both sexes (Bjørn Lomborg, *The skeptical environmentalist*); 2000–2015: Great Britain, both sexes (World bank).

The rate of increase in the twentieth century is impressive, and the fact that this trend is still continuing is significant. Extrapolating the trend from 1960 to 2015 in a linear way up to 2050 and 2100, the values obtained for Italy (where life expectancy is slightly higher than in Great Britain as reported in Fig. 13.1, at 83.49 years in 2015 compared to 81.61) are 92.6 and 105.7 years respectively. This increase is somewhat disappointing when compared with certain predictions based on recent medical advances which suggest that life expectancy should exceed 120 years much sooner, but is based on a simple extrapolation of the data for the last 50 years and therefore refers to a situation in which no relevant changes occur.

Technological development, particularly progress in the medical field, caused a very sharp rise in the population (Fig. 13.2; note the logarithmic scale). The figure, obtained using data provided by the United Nations Department of Economic and Social Affairs Population Division, for the years since 1950, and other sources for previous ones (with considerable uncertainty regarding ancient times), clearly shows that the increase has accelerated significantly since 1800 [3]. While the estimated figure for the population in 5000 BC (immediately after the Neolithic Revolution) was five million people, and at the beginning of the eighth century had reached about one billion, by 2000 it had exceeded six billion (to be precise 6,145,006,989). Currently (2019), the world population has exceeded 7.7 billion.

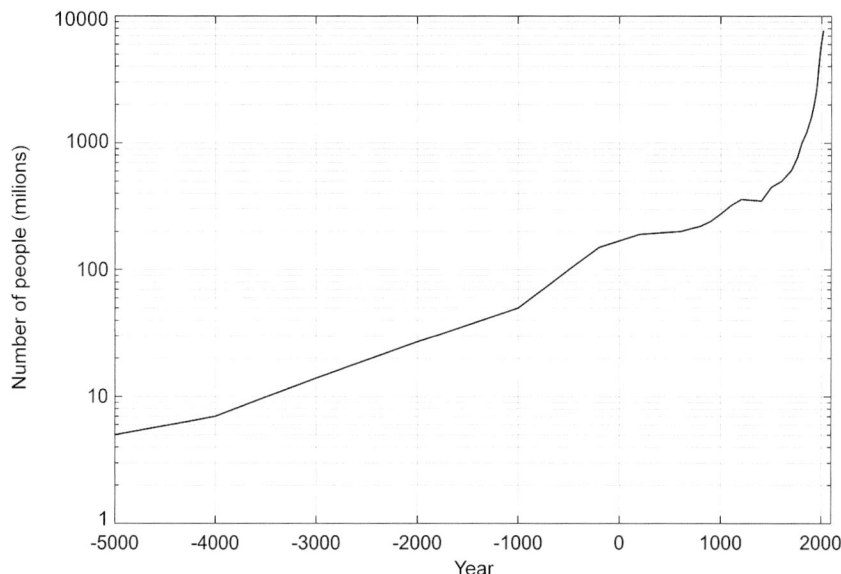

Figure 13.2. World population, from the Neolithic to present day.

The very rapid increase observed in the twentieth century would seem to confirm the most pessimistic Malthusian forecasts. However, the growth rate has recently decreased, from a maximum value of 2.09 percent per annum, reached in 1968, to the current 1.07 percent per annum. The forecasts for the future – which are not consistent with each other and include great uncertainties – suggest a further slowdown in the growth rate which will eventually lead to a slight decrease starting from the middle of this century, from a maximum population of about 11 or 12 billion.

Finally, it should be noted that from the nineteenth century, and particularly from the mid-twentieth century, the living conditions (at least the material ones) for the a rapidly growing population have improved, judging from the decrease, in relative terms and then also in absolute terms, of the number of people living in extreme poverty.

The population trends over the last two centuries, together with the number of people living in conditions of extreme poverty (poverty is not easily defined; in this case it was considered to be people with an average income of less than 1.9 equivalent dollars per day – a standard commonly accepted by the UN) are shown in Figure 13.3. The figures were reached by reporting the values obtained from the Our World in Data website [4]. This used data from *Inequality Among World Citizens 1820-1992* [5].

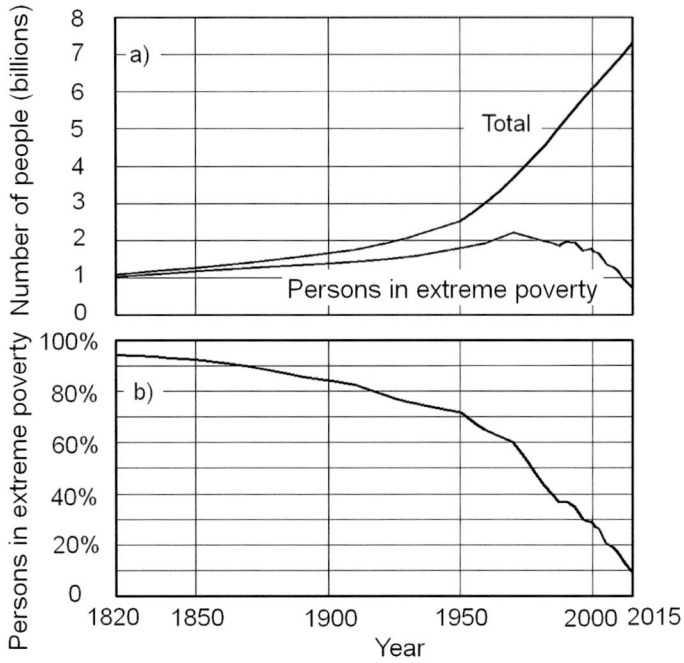

Figure 13.3 a) World population, number of people in extreme poverty in the last two centuries; b) Percentage of world population in conditions of extreme poverty in the same period.

Note that, at the beginning of the nineteenth century, more than 90 percent of the population was in extreme poverty (since the definition of poverty used here takes into account the effects of inflation, this very high value is quite reliable), while currently that figure has fallen to ten percent and is continuing to decrease. Eradicating poverty by 2030 is one of the primary objectives of the United Nations. The fact that living conditions are noticeably improving, despite the marked increase in population, is a radical confutation of the Malthusian approach, which, as we said in Chapter 10, did not take into account the effects of technological development.

There can be no doubt, from a material point of view, that it was worthwhile to develop an increasingly complex and widespread technology, and we must continue on the same path.

But can we reduce everything to a mere quantitative calculation; the number of goods we own and the length of our lives? With regard to the second aspect, for example, modern medicine is often accused of being able to lengthen our lives without worrying about the quality of that life. We live longer, but it seems that the years that have been added to our lifespan merely extend our old age, with all its problems. In fact, this is not accurate. Today, on average, we age more slowly and later on, while quality of life is at the center of medical research, alongside simply trying to ensure that we survive to a more advanced age.

Going beyond this purely quantitative perspective, Harari, in his previously cited book, wonders whether modern humans are happier than their prehistoric ancestors. But are we sure that this question makes sense? The first problem with this approach is linked to the fact that we are still unable to define what happiness is, more than 2000 years after philosophers like Aristotle and Epicurus began trying to do so. Equally, are we sure it makes sense to use happiness as a yardstick to measure the success of a society or a civilization?

Happiness was listed among the goals of a society for the first time in the Declaration of Independence of the United States, as an inalienable human right alongside life and freedom. Note that, according to Benjamin Franklin, the right in question was not happiness itself but the freedom to pursue it, which seems very wise since, like today, there was no agreement on how to define happiness at that time. It appears that Franklin was convinced to include the search for happiness among human rights by the Neapolitan philosopher Gaetano Filangeri who, in 1760, wrote "*In the concrete progress of the system of laws is the progress of national happiness, whose achievement is the true end of the government, which pursues it not generically but as the sum of the happiness of individuals*". It should be noted, however, that Filangeri did not refer to individual happiness, but to a sort of average happiness. All this is typical of the Enlightenment, and indeed the Masonic, way of thinking and would not have even been conceivable before that well-defined historical period.

Umberto Eco, in one of his writings on '*Il Corriere della Sera*' of March 26, 2014, suggested that "*...many of the problems that afflict us – I mean the crisis of values, the yielding to advertising seductions, the need to be seen on TV, the loss of our historical and individual memory... are due to this unhappy formulation of the American Declaration of Independence...*"

In essence, this insistence on happiness can be seen as both the cause and effect of exasperated individualism, the relativism of modern Western culture, and the current prevalence of rights over duties.

This approach from the American Declaration of Independence then passed into the UN declaration on human rights, which in 2012 even established the *International Day of Happiness* – which occurs on March 20 each year – and declares as a proven thing that the search for happiness is one of the fundamental purposes of humanity.

That being the case, it would seem that in order to answer to the question of whether it was worthwhile pursuing technology, we need to ask whether a Neolithic farmer was happier than a paleolithic hunter-gatherer (Filangeri would have said to ask whether the sum of the happiness of Neolithic farmers was greater than that of hunter-gatherers), or whether a miner who might be digging platinum on an asteroid in about thirty years will be happier than a computer technician currently working for Google or another large company.

Certainly, Augustine of Hippo would have set the problem in a completely different way. Since *"God created us for Him and our heart is restless until it can find rest in Him"*, happiness is a variable independent of everything else and depends only on our relationship with God [6]. Technology, of which we know Augustine was an admirer, therefore has nothing to do with happiness and so the question itself makes no sense.

Finally, Augustine would tell us that it was certainly worthwhile to develop our technology, even if our relationship with God remains the priority, because God has given us the task of taking care of, and enhancing, His creation. Technology allows us to improve the human condition, and this too is a divine commandment.

But Augustine, as one of the fathers of the church, was not the only one to come to the conclusions which lead us to believe that happiness has little to do with the technological developments of the society in which we live. On this point, most philosophers would agree with him. Happiness is something residing inside us and has very little to do with what we own or with our standards of living. Or, rather, if on the one hand living in poor material conditions can make it more difficult to reach happiness, there is no guarantee that living in comfort and in the (relative) security of a highly technological environment can reach it either.

So, if we cannot use happiness as a yardstick to measure the success of technological development, we simply have to go back to more measurable indicators, such as material success (number of people we can support, per capita income, total volume of the economy, etc.) and life expectancy. Here there are no doubts, because as soon as we use measurable parameters, the success of technological development becomes unequivocal.

But it is worth returning to a point already touched upon. When humankind started along the road leading to technology, it did so without realizing what it was doing and only really started asking questions about it recently. By then, our species had already started along a path from which it could not go back. Even as far back as the Neolithic Revolution, renouncing technology and economic development, or even reducing human involvement in them, would have been a huge disaster. It would not even have been possible to feed the large number of human beings that were inhabiting the areas of the planet which had taken up the new way of life.

Today, that progress has really picked up the pace and we are increasingly dependent on technology. Even in those cases in which technology itself creates problems, their solution can only be found in further development. In general, the various technologies make progress through the accumulation of knowledge and skills, and it is only through this accumulation that the problems created by insufficiently developed technologies can be solved.

A typical example can be seen in the automotive field. Right from the 1960s, when the problem of pollution due to internal combustion engines was raised, research, driven by increasingly restrictive legislation, has allowed us to produce engines that cause less and less pollution, albeit at the cost of increasing complexity and higher prices. The emissions from modern engines are a small fraction of those of the engines of about forty years ago, and the goal of almost zero-emission engines is now much closer thanks to radical innovations such as partial electrification (hybrid vehicles) and a large reduction in fuel consumption.

Problems are generally caused by insufficient technology and are solved with improvements, and in some cases even radical changes, in technology, which often require very large investments made possible only by an expansion of the economic base of society.

The real problem, if anything, is much different: Will we be able to maintain the pace of development we have witnessed over the last few centuries and in particular in the last few decades?

There are those who maintain that scientific development itself has limitations, and even that we are living in an age when science is about to reach those limits [7]. Following this line, it has been hypothesized that the current phase of rapid technological progress will end within a couple of centuries, possibly much earlier. However, these arguments are not very convincing. No-one has said that scientific and technological progress proceeds at a steady pace, and history shows that periods of rapid advancement have alternated with periods of stagnation, but this does not mean that any present hypothetical slowdown will lead to an overall stagnation of progress.

It is of course possible that there actually are limitations to technological progress. If the human species begins to extend into space, first in the solar system and then in other systems, the huge distances between the settlement centers and the limited speed at which they can be reached, according to modern physics, will necessarily imply a significant slowdown of development and, most likely, a stabilization.

It is also possible that the extension of human lifespan which some people predict for the future will materialize and humans will live much longer than today. Actually, it seems that the aging process may be controllable, at least to a certain extent, even if there is no clear idea how far human life span could be extended. There are even those who say that a sort of immortality could be achieved at some point in the distant future, at least in the sense of an indefinite lengthening of life. As we have already seen, the hope that scientific progress would lead to this outcome was expressed at the end of 1700 by Vincenzo Monti in his *Ode al signor di Montgolfier*.

Even in the most optimistic forecasts, however, this will be only a relative immortality, or better, 'amortality', given that accidental death cannot be eliminated. A significant increase in lifespan must necessarily be accompanied by a noticeable drop in birth rate and a slower generational change, which in turn would probably cause a slowdown in innovation in all fields. It would certainly make it difficult to produce the scientific revolutions which, up to now, have always required the replacement of one generation of scientists with the next.

These ideas concerning the end of progress and technological stagnation realistically derive from our difficulty in imagining life in a context radically different from the current one. Certainly, a Cro-Magnon man could never have imagined life in a village of Neolithic farmers, while just two centuries ago, the claim that only a small minority of people would eventually be engaged in cultivating the land would have raised a great concern about what everyone else could possibly do.

Each generation has always had the tendency to think that the apex of human development had just been reached. A typical example would be the statements concerning the end of science that circulated among physicists in the last part of the nineteenth century, when the belief that all that could be discovered and invented had already been achieved was so strong that many scientists discouraged their best students from undertaking a scientific career.

However, it is likely that these are not objective limitations to the progress of science and technology. If technological developments cease in the future and if the promises we saw in Chapter 12 remain only on paper, it will likely be for reasons of a completely different kind, similar to those we have discussed regarding the end of Hellenistic science and technology.

A substantial part of Western civilization has recently lost the confidence it placed in itself and in its founding values, and above all confidence in progress and technology. A cultural relativism is spreading that endangers the very foundations on which the West has built its scientific and technological development, and thus also its industrial and economic development.

The idea that Western civilization had started an unstoppable decline began to spread from the beginning of the twentieth century and, in particular, after the end of World War I. One of the major contributions to this idea came from the philosopher and historian Oswald Spengler with his *The Decline of the West*, published in 1918 (the second volume appeared in 1923) [8]. Since he started writing it about ten years earlier, his pessimistic vision was therefore not the result of the war and its particularly dire consequences for Germany. The essay *The Course of Civilizations*, by Johan Huizinga, printed in 1936, followed similar themes [9].

It must be said that pessimistic visions about the future of civilization are not new: Bjøn Lomborg quoted an ancient Assyrian tablet that reads: "*Our land is degenerating in these times, corruption spreads, children do not obey their parents; all men want to write a book, and the end of the world is evidently approaching*" [10, 11]. Although the original citation does not give precise references (and therefore the whole quotation may be apocryphal), this sentence is representative of a number of statements that can be found in all cultures throughout history. Ultimately, it is nothing but another re-edition of the myth of the 'Golden Age'. Perhaps it is better suited as study material for psychologists and sociologists than as a prediction about the future of civilization (any civilization).

The fact is that the impression of living in a phase of decline of civilization is now widespread in the West. It has spread among people of all political beliefs, from conservatives (who dream of a past that they were unable to preserve) to people who consider themselves as progressive (even if they no longer believe in 'progress', as the proponents of nineteenth century scientific and technological progress would have defined it). There is no doubt that Umberto Eco was one of latter, from the previously quoted passage: "*the crisis of values, the surrender to the advertising seductions, the need to be seen on TV, the loss of historical and individual memory*". Or the astrophysicist Giovanni Bignami who, at the end of his book *Le rivoluzioni dell'universo* spoke about the future of *homo sapiens* [12]. He mentioned the promises of improvement in our species thanks to the gene therapy and other technologies mentioned in Chapter 12, but also spoke of a widening cultural gap between the different components of our society, of a mass lack of culture that can be defined as illiteracy, which he stigmatizes as: "*…a large percentage of the millennial generation no longer knows how to write, in the elementary sense of correctly making a dictation, either with paper and pencil or with a keyboard, or read in the sense of reading aloud (and understanding) a random page of a newspaper, not to mention the ability to express original or, worse, abstract concepts, as with many of the older people, who do not perform better.*" He then talks about a revolution of ignorance.

This ignorance is primarily a scientific ignorance, linked to the inability to distinguish sciences from pseudo-sciences, as evidenced by the success of astrology and alternative medicines, including homeopathy, parapsychology, the spread of conspiracy theories, and a thousand other pseudo-sciences, greatly facilitated by the uncontrollable distribution of a network in which any idea can be spread without any sort of filter.

The chances of continuing on our way to a future in which the promises of technology will come about depends on the possibility of overcoming the wave of irrationalism and ignorance that seems to be overwhelming us; to resume, on the basis of our roots that start from the Greek and then the Judeo-Christian civilization, the path of rationality that brought us to where we are. Above all, this is what we owe to other cultures that, in the global perspective, have begun that path of modernization and Westernization that Huntington spoke about. Such Westernization does not mean subjugation to the West, but simply making use of the 'dogma' on which Western civilization was based: The world is rational, and human reason is the instrument that allows it to be understood.

References

1. R. Dawkins, *The selfish gene*, Oxford University Press, Oxford 1976.
2. B. Lomborg, *The skeptical environmentalist*, Cambridge University Press, Cambridge 1998.
3. http://www.worldometers.info/world-population/world-population-by-year/
4. https://commons.wikimedia.org/w/index.php?search=world+population+in+extreme+poverty&title=Special%3ASearch&profile=advanced&fulltext=1&advancedSearch-current=%7B%22namespaces%22%3A%5B6%2C12%2C14%2C100%2C106%2C0%5D%7D&ns6=1&ns12=1&ns14=1&ns100=1&ns106=1&ns0=1#/media/File:World-population-in-extreme-poverty-absolute.svg and https://commons.wikimedia.org/w/index.php?search=world+population+in+extreme+poverty&title=Special%3ASearch&profile=advanced&fulltext=1&advancedSearch-current=%7B%22namespaces%22%3A%5B6%2C12%2C14%2C100%2C106%2C0%5D%7D&ns6=1&ns12=1&ns14=1&ns100=1&ns106=1&ns0=1#/media/File:World_population_living_in_extreme_poverty_-_Our_World_in_Data_-_2015.png
5. Bourguignon and Morrisson (2002) – *Inequality Among World Citizens: 1820–1992*. In American Economic Review, 92, 4, 727–748
6. *Fecisti* nos ad te et inquietum est *cor* nostrum, donec requiescat in te. (Confessions, I, 1).
7. J. Horgan, *The End of Science, Facing the Limits of Knowledge in the Twilight of the Scientific Age,* Basic books, New York, 1996, P. Musso, *How Advanced is ET,* VII Trieste Conference on Chemical Evolution and the Origin of Life: Life in the Universe, Trieste, Settembre 2003.
8. O. Spengler, *The Decline of the West*, Alfred A. Knopf, New York 1962.
9. J. Huizinga, *The Course of Civilizations*, Cluny media, 2018.
10. Bjøn Lomborg, T*he Skeptical Environmentalist,* cit.
11. J. Simon, *The ultimate Resource 2*, Princeton University Press, Princeton 1996.
12. Giovanni Bignami, *Le rivoluzioni dell'universo*, Firenze, Giunti, 2017.

Appendix: Chronological tables

Table A.1. Some of the philosophers, scientists, writers, rulers and founders of religions in Greek and Hellenistic times (up to the ousting of the Greeks from Alexandria). The birth dates and even some death dates are uncertain and only an estimate is provided.

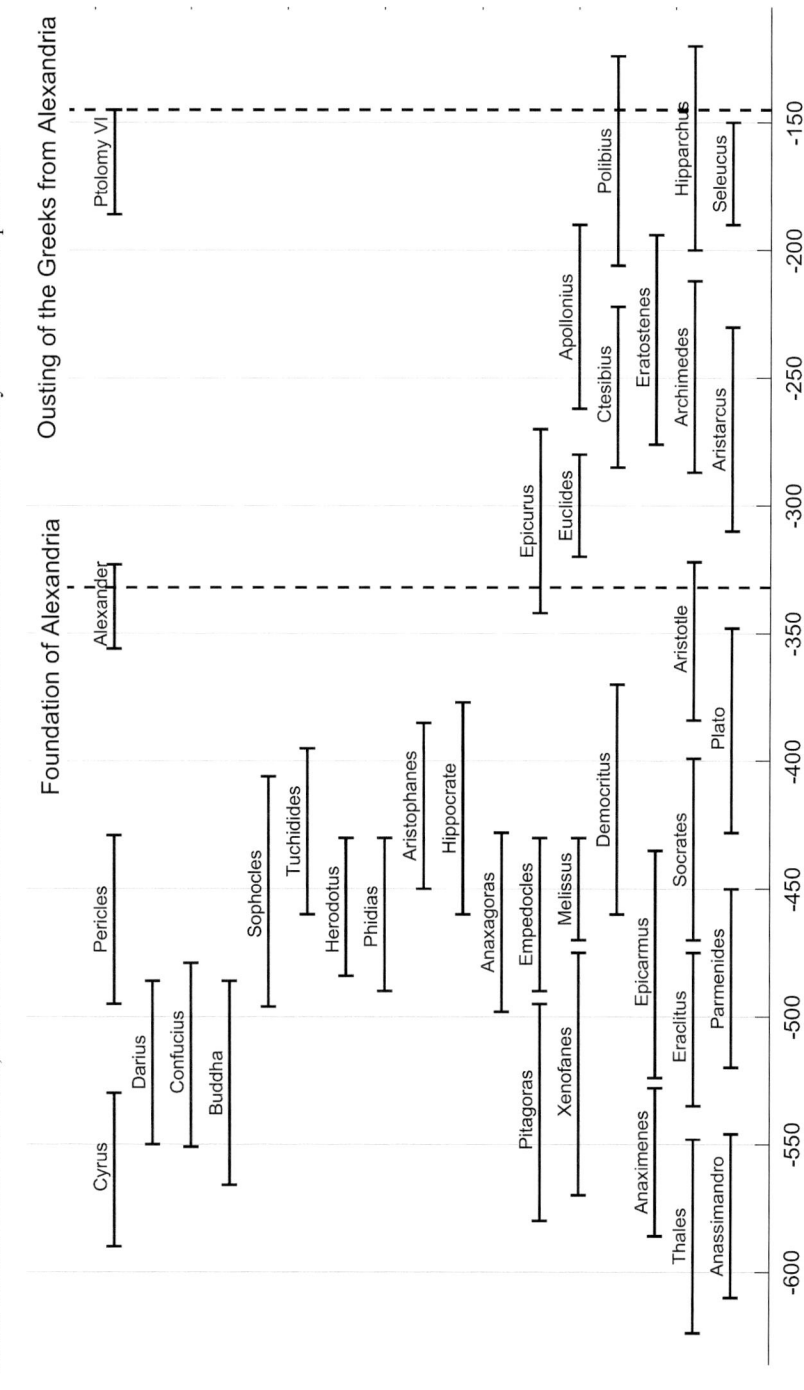

252 Appendix: Chronological tables

Table A.2. Some of the philosophers, scientists, writers, rulers and founders of religions in Roman times. The birth dates and even some death dates are uncertain and only an estimate is provided.

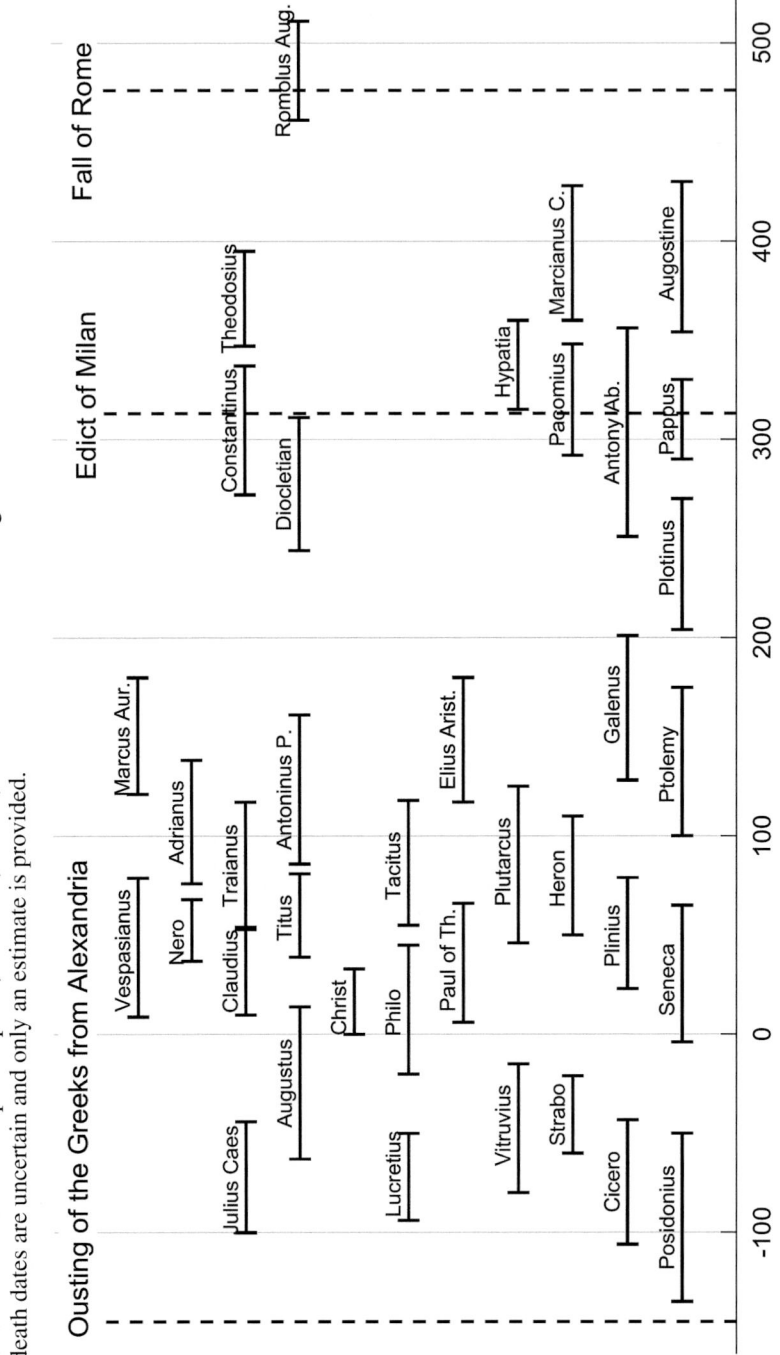

Appendix: Chronological tables 253

Table A.3 Some of the philosophers, scientists, writers, rulers and founders of religions in the Middle Ages.

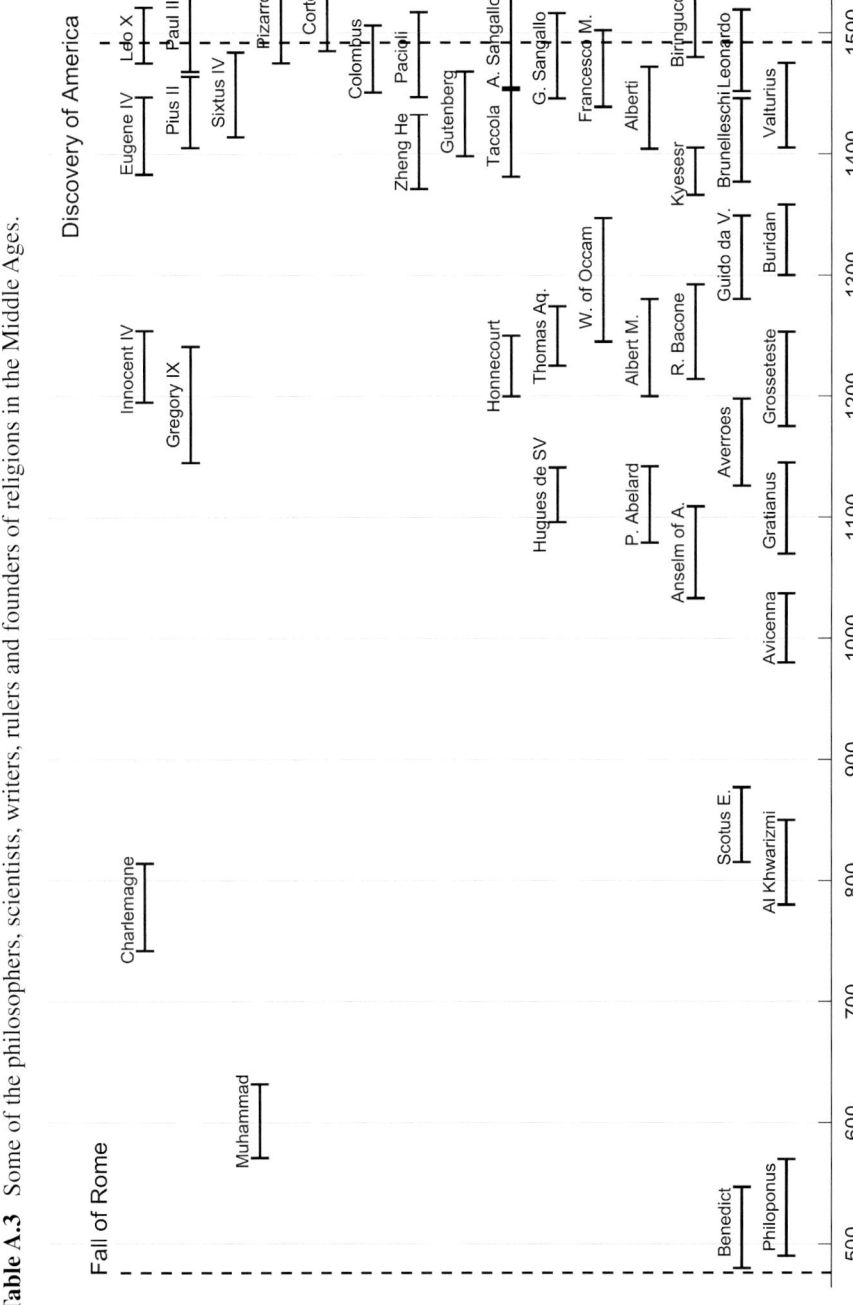

254   Appendix: Chronological tables

Table A.4 Some of the philosophers, scientists, writers, rulers and founders of religions in the Modern Age.

Bibliography

1. Abbagnano N., *Storia della filosofia, vol. I: Filosofia antica. Filosofia patristica. Filosofia scolastica,* Utet, Turin, Italy 1946
2. Abbagnano N., *Le sorgenti irrazionali del pensiero,* Marte, Salerno 2008
3. Abgrall J. M., *Healing or stealing? Medical charlatans in the new age,* Algora, New York 2000
4. Alleau R., *Hitler et les sociétés secrètes, enquête sur les sources occultes du nazisme,* Cercle du nouveau livre d'histoire, Paris 1969
5. Antinori C., *La contabilità pratica prima di Luca Pacioli: origine della partita doppia* in De Computis, 1 (2004), pp. 4–23
6. Artifoni E., *Storia medievale,* Donzelli, Rome 2003
7. Babich B., *Between Heidegger and Adorno: airplanes, radio and Sloterdijk's terrorism* in "Kronos Philosophical Journal", vol. VI, 2017, pp. 133–158
8. Barber M., *The Trial of the Templars,* Cambridge University Press, Cambridge 1978
9. Benham A., *English Literature from Widsith to the Death of Chaucer,* Yale University Press, New Haven 1916
10. Bennett F., *Artefactualising the sacred: restating the case for Martin Heidegger's 'Hermeneutical' philosophy of technology* in C. Deane-Drummond, S. Bergmann e B. Szersynsky (a cura di), "Technofutures, Nature and the Sacred: transdisciplinary perspectives", Ashgate, Farnham 2015
11. Bignami G., *Le rivoluzioni dell'universo,* Giunti, Firenze 2017
12. Bignami G., *Oro dagli asteroidi e asparagi da Marte,* Mondadori, Milan, Italy 2015
13. Bishop M., Green M., *Face it: Money is Technology, and we can do better than gold,* in "Business Insider", online, 2013
14. Boccaccini G., *Oltre l'ipotesi essenica: lo scisma tra Qumran e il giudaismo enochico,* Morcelliana, Brescia 2003
15. Bourguignon F., Morrisson C., *Inequality Among World Citizens: 1820–1992.* In American Economic Review, vol. 92, n. 4, 2002, pp. 727–748
16. Bridbury A. R., *The dark ages,* in "The Economic History Review", n. 22, 1969, pp. 526–537
17. Cambiano G., *Platone e le tecniche,* Laterza, Rome-Bari, Italy 1991
18. Cambiano G., *Storia della filosofia antica,* Laterza, Rome-Bari, Italy 2004

19. Capurro R., *Le positivisme est un culte des morts: Auguste Comte*, Epel, Paris 1998
20. Cassirer E., *The philosophy of the enlightenment*, Beacon Press, Boston 1955
21. Chianazzi P., *Gli ordini cavallereschi: storie di confraternite militari*, Edizioni Universitarie Romane, Rome 2013
22. Ciambotti M., *Finalità e funzioni della contabilità in partita doppia nell'opera di Luca Pacioli*, in F.M. Cesaroni, M. Ciambotti, E. Gamba, V. Montebelli, *Le tre facce del poliedrico Luca Pacioli*, Quaderni del Centro Internazionale di Studi, Urbino e la Prospettiva, Arti Grafiche Editoriali, Urbino 2010, pp. 11-25
23. Cipolla C. M., *Clocks and Culture (1300–1700)*, W. W. Norton & Company, London 1967
24. Clendinnen I., *Aztecs: an interpretation*, Cambridge University Press, Cambridge 1995
25. Coccia M., *Le origini dell'economia dell'innovazione: il contributo di Rae*, Ceris-Cnr: Istituto di Ricerca sull'Impresa e lo Sviluppo, working paper 1/2004, pp.1–19
26. Dawkins R., *The selfish gene*, Oxford University Press, Oxford 1976
27. Dawson C., *The Making of Europe: An Introduction to the History of European Unity*, Sheed and Ward, London 1932, reissued by the Catholic University of America Press, Washington DC 2003
28. Dewar J., *The Nuclear Rocket*, Apogee Books, Burlington, Canada 2009
29. Diamond J., *Guns, Germs and Steel: The Fates of Human Societies*, W.W. Norton, New York 1997
30. Drexler K. E., *Engines of Creation, The Coming Era of Nanotechnology*, Anchor Books, New York 1986
31. Durschmied E., *Unsung Heroes: The Twentieth Century's Forgotten History-Makers*, Hodder & Stoughton, London 2003
32. Eco U., *La bustina di Minerva*, "L'Espresso", Roma, 26 marzo 2014
33. Eliade M., *Occultismo, stregoneria e mode culturali: saggi di religioni comparate*, Lindau, Turin, Italy 2018
34. Farrington B., *Philosophies of Technology: Francis Bacon and his Contemporaries*, Liverpool University Press, Liverpool 1964
35. Feenberg A., *Heidegger and Marcuse: The Catastrophe and Redemption of History*, Routledge, London 2005
36. Finney B., *From Sea to Space*, The Macmillan Brown Lectures, Massey University, Hawaii Maritime Center 1992
37. Finney B., *The Prince and the Eunuch, in Interstellar Migration and the Human experience*, University of California Press, Berkeley 1985
38. Galli G., *Hitler e il nazismo magico*, Rizzoli, Milan, Italy 2005
39. Gargano A., *L'irrazionalismo dell'Ottocento*, La Città del Sole, 2005
40. Gibbon E., *The History of the Decline and Fall of the Roman Empire*, 6 volumes, Strahan & Cadell, London 1779-1796
41. Giusto N., *Heidegger, Gestell e la questione della tecnica* in "Culture Digitali" (online), 2015
42. Goody J., *The Domestication of the Savage Mind*, Cambridge University Press, Cambridge 1977
43. Green M., *In gold we trust? The future of money in an age of uncertainty*, ebook, The Economist Newspaper Limited 2013
44. Guardini R., *Lettere dal Lago di Como: La tecnica e l'uomo (1923–25)*, Morcelliana, Brescia 1993.
45. Harari Y. N., *21 Lessons for the 21st Century*, Spiegel & Grau, New York 2018
46. Harari Y. N., *Homo Deus: A Brief History of Tomorrow*, Harvill Secker, London 2016

47. Harari Y. N., *Sapiens. From Animals into Gods: A Brief History of Humankind*, Harper, New York 2015
48. Harmless W., *Desert Christians. An Introduction to the Literature of Early Monasticism*, Oxford University Press, Oxford 2004
49. Harris W. V., *Ancient Literacy*, Harvard University Press, Cambridge (Massachusetts) 1989
50. Heppenheimer T. A., *A Brief History of Flight*, Wiley, New York 2001
51. Horgan J., *The end of science, facing the limits of knowledge in the twilight of the scientific age,* Basic books, New York, 1996
52. Huizinga J., *The Course of Civilizations*, Cluny Media, Providence 2018
53. S. P. Huntington, *The clash of civilizations and the remaking of the world order*, Simon & Schuster, New York 1996
54. Hutin S., *A History of Alchemy – Ancient Science of Changing Common Metals Into Gold*, Tower Publishing, New York City 1962
55. Irokawa D., The *Age of Hirohito, in search of Modern Japan,* Free Press, 1995
56. Jaynes J., *The Origin of Consciousness in the Breakdown of the Bicameral Mind*, Houghton Mifflin, Boston and New York, 1976
57. Johnson L., *Sky Alert!*, Springer, Chichester 2013
58. Kelly J., *The Great Mortality: An Intimate History of the Black Death, the Most Devastating Plague of All Time*, Harper, New York 2005
59. Koyré A., *From the Closed World to the Infinite Universe*, Johns Hopkins Press, Baltimore 1957
60. Kurzweil R., *The singularity is near*, Viking Press, New York 2005
61. Lloyd G. E. R., *Early Greek Science: Thales to Aristotle*, W. W. Norton & Company, New York, 1970.
62. Lomborg B., *The skeptical environmentalist*, Cambridge University Press, Cambridge 1998.
63. Long A. A., *Hellenistic Philosophy Stoics, Epicureans, Sceptics*, Bristol Classical Press, Bristol 1974
64. Manchester W., *A world lit only by fire: the medieval mind and the Renaissance, portrait of an age*, Little Brown & C., New York 1993
65. Mankins J. C., ed., *Space solar power*, International Academy of Astronautic, Paris 2011
66. Marks J., *An Amoral Manifesto* in *Philosophy Now*, online, 2010
67. Martin F., *Money: The Unauthorized Biography – From Coinage to Cryptocurrencies*, Bodley Head, London 2013
68. Martone C., *Il Giudaismo antico*, Carocci, Rome 2008
69. Mazzinghi L., *Il libro della Sapienza: elementi culturali in* "Ricerche storico-bibliche", 10, 1998, p. 179–198
70. Meadows D.H., Meadows D.L., Randers J., Behrens W.W., *Limits to Growth, 1972.*
71. Morel J. P., *L'artigiano*, Laterza, Rome-Bari, Italy 2013
72. Morowitz H.J., *The emergence of everything*, Oxford University Press, Oxford 2002
73. Musso P., *How Advanced is ET,* VII Trieste Conference on Chemical Evolution and the Origin of Life: Life in the Universe, Trieste, September 2003
74. Napoleoni C., *Smith Ricardo Marx: considerazioni sulla storia del pensiero economico*, Boringhieri, Turin, Italy 1970
75. Noble D. F., *The religion of technology: the divinity of man and the spirit of invention*, New York and London 1997
76. Oden T.J., Bathe K.J., *A Commentary on Computational Mechanics*, Applied Mechanics Review, 31, 1978.

77. Osborne R., *Civilization: a new history of the Western world*, Pegasus, New York 2006
78. Pekàry T., *Die Wirtschaft der griechisch-römischen Antike*, Steiner, Wiesbaden 1979
79. K. Popper, *The Open Society and Its Enemies*, Routledge, London 1945.
80. Rapetti A., *Storia del monachesimo occidentale*, Il Mulino, Bologna, Italy 2013
81. Ratzinger J., Pera M., *Senza radici. Europa, relativismo, Cristianesimo, Islam*, Mondadori, Milan, Italy 2004
82. Renzi L., Andreose A., *Manuale di linguistica e filologia romanza*, Il Mulino, Bologna, Italy 2003
83. Righetto G., La scimmia aggiunta, Paravia, Turin, Italy 2000
84. Rostovzev M. I., *The Social and Economic History of the Hellenistic World,* Clarendon Press, Oxford 1941
85. Russell B., Foulkes P. (ed.), *Wisdom of the West,* Macdonald, London 1959
86. Russo L., *Flussi e riflussi*, Feltrinelli, Milan, Italy 2003
87. Russo L., *The forgotten revolution*, Springer, Berlin 2004
88. Sacchi P., *The History of the Second Temple Period*, T. & T. Clark, Edinburgh 2004
89. Salomon W., *The Economics of Interstellar Commerce*, in *Islands in the Sky*, Wiley, New York 1996
90. Saracco G., *Chimica Verde 2.0*, Zanichelli, Bologna, Italy 2017
91. Schiavone A., *The End of the Past: Ancient Rome and the Modern West*, Harvard University Press, Harvard 2000
92. *Schiavone A., Spartacus in* "Revealing antiquity" n. 19, *Cambridge, MA 2011*
93. Simon J., *The ultimate Resource 2*, Princeton University Press, Princeton 1996
94. Smith M., *Palestinian Judaism in the first century* in M. Davis, *Israel: its role in civilization*, Harper and Brothers, New York 1956
95. Sokal A., Bricmont J., *Intellectual Impostures,* Picador, London 1999,
96. Sommariva A., Bignami G., *The Future of Human Space Exploration, Macmillan*, London 2016
97. Spengler O., *The Decline of the West*, Alfred A. Knopf, New York 1962
98. Stark R., *Bearing False Witness: Debunking Centuries of Anti-Catholic History*, Templeton Press, West Conshohocken 2016
99. Stark R., *How the West Won. The Neglected Story of the Triumph of Modernity*, Intercollegiate Studies Institute, Wilmington, Delaware 2014
100. Stark R., *One True God Historical Consequences of Monotheism,* Princeton Unversity Press, Princeton 2001
101. Stark R., *The Triumph of Christianity: How the Jesus Movement Became the World's Largest Religion,* Harper and Row, San Francisco 1970
102. Tanzella Nitti G., *Pensare la tecnologia in prospettiva teologica: esiste un umanesimo scientifico?*, in *Scienza, tecnologia e valori morali: quale futuro?* Studi in onore di Francesco Barone, a cura di P. Barrotta, G.O. Longo, M. Negrotti, Armando, Rome 2011, pp. 201–220
103. Tanzella Nitti G., *Progresso scientifico e progresso umano. Prospettive filosofiche e teologiche*, in S. Rondinara (ed.), *La scienza tra arte, comunicazione e progresso*, Città Nuova, Rome 2017, 265–283
104. Tanzella Nitti G., *Teologia fondamentale in contesto scientifico,* Città Nuova, Rome 2018
105. Teilhard de Chardin P., *The Phenomenon of Man*, Harper. New York 1959
106. Timossi R. G., *Nel segno del nulla*, Lindau, Turin, Italy 2016
107. Valori G. E., *Spiritualità e illuminismo*, Futura, Perugia 2018
108. Vione D., *Perché l'omeopatia non è una scienza*, www.cicap.it, 2004

109. Walbank F.W., *The Decline of the Roman Empire in the West*, University of Toronto, Toronto 1969
110. Weber M., *Economy and society*, University of California Press, Oakland 1978.
111. Weeber K.W., *Smog über Attika*, Artemis Verlag Zürich and München 1990
112. Wells P. S., *Barbarians to Angels: The Dark Ages Reconsidered*, W.W. Norton, New York 2008
113. White L. Jr., *Cultural Climates and Technological Advance in the Middle* in "Viator" 2 (1971), pp.171–202.
114. Wilkinson T., *The rise and fall of Ancient Egypt*, Random House, New York 2010
115. Young, J., *Heidegger's Later Philosophy*, Cambridge University Press, Cambridge 2002
116. Zellini P., *Numero e Logos*, Adelphi, Milan, Italy 2010

Index

A

Accounting, 118–123, 161
Adam, 60, 69, 71, 110, 112, 162, 167, 170–172
Additive manufacturing, 202
Adorno (Theodor), 220
Aelia Pulcheria, 106
Aerodynamics, 133, 139, 141, 143, 144, 149
Age of Saturn, 69
Agricultural revolution, 183–185
Agriculture, 14, 22, 29, 30, 54, 60, 72, 74, 85, 93, 94, 112, 117, 126, 162, 173, 183, 184, 189, 190, 199, 200, 224, 241
Air pollution, 190, 227
Alaric, 101
Al-Battani (Albategnius), 105
Alberti (Leon Battista), 128
Alchemy, 52, 109, 207, 208
Alexander the Great, 47, 58, 61, 86, 87
Aliatte, 37
Ali Pascia, 203
Al-Khwārizmī, 104, 105
Al Mamun, 105
Al Mutawakkil, 105
Ampère (André-Marie), 143
Anaxagoras, 41, 43, 57, 58, 217
Anaximander, 41
Anaximenes, 41, 58
Anselm of Aosta, 210
Anthroposophy, 211, 216, 217
Antinori (Carlo), 122
Antoninus Pius, 77
Antony, 58, 114, 115

Apes, 2–4, 8, 11
Apocryphal gospels, 74, 210, 211, 213
Apollodoro of Damascus, 34
Archimedes, 33, 49–53, 80, 131, 133, 137, 181
Archimedes' screw, 50
Aristarchus, 48
Aristide (Elio), 77, 78, 83, 103
Aristophanes, 31
Aristotle, 31, 32, 45, 47, 48, 51, 58, 65, 72, 104, 107–109, 120, 217, 245
Art, 13, 20, 31, 32, 42, 74, 80, 84, 111–113, 117, 120, 128, 131, 170, 206, 220
Artificial
  intelligence, 228, 230
  satellite, 195
  selection, 15, 21
Artifoni (Enrico), 103
Asset bubbles, 160
Astrology, 52, 53, 208, 249
Atheism, 214–216
Athena, 42
Atlantis, 69, 217
Atmosphere, 10, 180, 201, 226, 227, 234, 235
Atmospheric pressure, 134, 135, 137, 149
Attila, 101
Augustine, 75, 105, 110, 111, 172, 217, 246
Augustine of Hippo, 74, 246
Augustus, 81, 84
Australopithecus, 2
Automobile, 192, 193, 203
Autotroph, 14
Avicenna, 104, 108

Index

B
Bacon, F., 126, 133
Bacteriological warfare, 175, 176
Ballista, 86, 127
Banks, 31, 32, 55, 155, 156, 159, 160, 164, 175, 216, 242
Barber (Richard), 121
Bartering, 35
Bathe (Klaus-Jürgen), 197
Bauer (Georg, Agricola), 129
Bell (Alexander Graham), 144
Benedict of Nursia, 115
Benedict XVI, 223
Benz (Ernst), 115
Bergeron (L. E.), 131
Bernard of Chartres, 99
Berthollet (Claude-Louis), 151
Besson (Jacques), 129
Bezaleel, 60
Big Bang, 68
Big Crunch, 68
Big Data, 199, 230
Bignami (Giovanni), 249
Biofuels, 227
Biological computers, 231
Biosphere, 10, 234, 235
Bishop (Matthew), 35, 106, 107
Black (Joseph), 136
Black death, 176
Blavatsky, Madame (Eléna Petróvna), 213, 217
Bohr (Niels), 145
Bombing of Dresden, 175
Bonaventura di Bagnoregio, 210
Bottom-up manufacturing, 202
Boulton (Mattew), 136
Bow, 12, 13, 108
Brain, 2–4, 11, 14, 45, 231, 239
Breeding, 29, 54, 235
Brickmont (Jean), 198
Bridbury (Anthony Randolph), 103
British East India Company, 156
Bronze age, 15, 17, 18, 69, 241
Browning (John), 140
Brunelleschi (Filippo), 128
Bukhtishu, 105

C
Caesar (Julius), 73, 94, 174, 175
Cambiano (Giuseppe), 43, 45
Caminos (Ricardo), 30
Camus (Albert), 208
Cannibalism, 30
Capurro (Raquel), 215

Cardano (Girolamo), 143
Carnot (Lazare), 151
Carnot (Sadi), 136
Cart, 26, 27, 89, 93, 94
Cast iron, 18, 86, 186–188, 192
Categorical imperative, 169, 170
Cavendish (Henry), 138
Cayley (Sir George), 139–141
Ceramics, 17, 20, 22, 74, 93, 101
Cervantes (Miguel), 132
Chanute (Octave), 140, 141
Charlemagne, 111, 115
Charles (Jacques Alexandre César), 138
Charles the Bald, 99, 111
Charles V, 131, 155
Chateaubriand (François-René), 213, 214
Chaucer (Geoffrey), 99
Chesterton (Gilbert Keith), 213
Chimpanzees, 2, 137
Christianity, 70–75, 78, 90, 92, 102, 103, 105, 106, 110–113, 118, 159, 167, 210, 211, 213, 214, 216
Cicero, 51, 80, 83, 85, 93
Cipolla (Carlo), 131
Circular time, 66–70
Clarke (Bruce C.), 179
Clausius (Rudolph), 136
Clement IV, 109
Cleomenes III, 174
Climate change, 22, 185, 200
Clocks, 49, 126, 131, 132
Club of Rome, 185
Coal, 93, 136, 148, 186, 190, 192, 193, 200, 220, 224, 227
Cochlea, 50, 54
Cognitive revolution, 7–10, 205, 206, 239
Cold war, 179, 195, 204
Colombus (Cristopher), 154
Combine harvester, 93, 94
Communism, 167, 170
Compass, 48, 115, 126, 149, 203
Computers, 133, 180, 194, 197–199, 201, 202, 220, 228–231, 246
Comte (Auguste), 214
Consumerism, 194
Cook (James), 157
Copper age, 17
Cortés (Hernán), 153, 154, 166
Craftsman, 29, 31, 32, 34, 43, 45, 60, 65, 66, 71, 72, 80, 93, 110, 129, 132, 192
Crank mechanism, 91
Crossbow, 127, 173
Crowley (Aleister), 208
Ctesibius, 49, 50, 52

Culture, 5, 8–10, 14, 15, 21, 32, 39, 47, 52, 58, 61, 64, 74, 100–102, 104, 105, 112, 131, 151, 164, 167, 203, 204, 206, 211, 212, 245, 249
Cyborg, 230
Cyril (bishop), 106, 107

D

Daikichi Irokawa, 177
d'Alembert, 139, 143
Dante Alighieri, 53, 120
Darby (Abrham), 188
Darius, 44
Darwin (Charles), 157
Dawkins (Richard), 240
Dawson (Christopher H.), 105, 110
Declaration of Independence of the United States, 245
Defoe (Daniel), 185, 203
De Forest (Lee), 228
Deism, 214
de La Mettrie (Julien Offray), 215
Demiurge, 32, 45, 65, 66, 132
Democritus, 7, 43, 45
Demonology, 213
Dhimmi, 104, 105
Diamond (Jared), 14, 15, 173
Dietrich D'Holbach (Paul Heinrich), 215
Dionysus, 207
Djoser, 32
DNA, 19, 70–72, 75, 199, 200, 202, 231
3-D printing, 202
Dual technologies, 172–173, 180–182
Dutch East India Company, 156

E

Eco (Umberto), 245, 246
Edison (Thomas Alva), 144
Einstein (Albert), 145, 178
Eisenhower (Dwight David), 146, 179
Electrical industry, 18, 192
Electricity, 143–147, 149, 192, 219, 235
Electronics, 194, 198, 228–231
Elia, 69, 214, 215
Eliade (Mircea), 208
Elio Aristide, 77, 78, 83, 103
Empedocles, 41, 211
Energy, 11, 12, 15, 24, 66, 86, 90–93, 104, 135, 136, 144–150, 153, 173, 176, 179, 180, 185, 192, 193, 200, 205, 217, 219, 224–228, 230, 231, 233
Engineering schools, 87, 150, 215, 220
Enoch, 69

Epicharmus, 41, 42
Epicurus, 245
Epimetheus, 41–43
Euclid, 48, 52, 53
Eugene IV, 168
Eumenes II, 47
Evolution, 2, 3, 10, 12, 13, 19, 30, 41, 79, 82, 83, 99, 105, 110, 118, 122, 125, 153, 157, 158, 164, 165, 167, 193, 211, 215, 218, 226, 236, 237, 239, 241
Extraterrestrial resources, 233

F

Faraday (Michael), 143, 144
Farrington (Benjamin), 133
Fascism, 170
Feenberg (Andrew), 219
Fermi (Enrico), 145, 146
Fidia, 112
Filangeri (Gaetano), 245
Finney (Ben), 171, 237
Fire, 10–13, 20, 22, 42, 43, 45, 49, 67, 90, 92, 97, 148, 172, 174, 175, 180, 190, 206, 239, 241
Firearms, 127
Fireplace, 22, 126
Flavius Aetius, 101
Foljambe (Joseph), 183
Force, 7, 12, 23, 24, 27, 48, 51, 52, 81, 83, 99, 100, 103, 108, 109, 114, 115, 126, 127, 133, 138, 139, 141, 152., 153, 155, 167, 173, 177, 178, 181, 189, 190, 206, 211, 217
Fossil fuels, 136, 147, 196, 200, 224–226, 233
Fourth Industrial Revolution, 199, 200, 203, 223, 224, 228
Francesco di Giorgio Martini, 129
Franklin (Benjamin), 143, 245
Fulcanelli, 208

G

Galileo Galilei, 70, 107, 132–134
Galvani (Luigi), 143, 145
Gates (Bill), 235
Gear wheel, 49, 91, 129
Genetically modified organisms, 199, 200
Geocentrism, 53
Gibbon (Edward), 2, 97
Gibbons, 2, 97
Glasses, 125
Glass windows, 89, 126
Glendinning (Chellis), 220
Globalization, 78, 81, 164, 165, 199
Global village, 198

Gnosis, 210
Gnosticism, 114, 210, 213
Goddess Reason, 214
Golden Age, 69, 71, 78, 122, 132, 241, 249
Góngora (Luis), 132
Gorillas, 2
Gould (Stephan Jay), 236
Gracián (Baltasar), 132
Gratian, 119
Green (Michael), 35
Greenhouse gases, 185, 226, 227
Gregoire (Henry), 214
Gregory IX, 120
Guido da Vigevano, 128
Gunpowder, 39, 126, 127, 148, 149, 203
Gutenberg (Johannes), 125

H
Hahn (Otto), 145
Hammurabi, 30, 37
Harari (Yuval Noah), 7, 35, 152, 164, 239, 240, 245
Harmless (William), 114
Harris (William V.), 98
Heidegger (Martin), 218–220
Heliocentrism, 48
Hellenistic science, 48, 51–54, 133, 143, 248
Helvétius (Claude-Adrien), 215
Hephaestus, 31, 42
Heraclitus, 57, 58
Hermes Trismegistus, 207, 217
Hermeticism, 207
Herodotus, 40, 44
Heron, 33, 51, 90, 92, 134
Heterotroph, 14
Hieron, 51
Hipparchus, 48
Hippocrates, 45, 46, 112
Hirohito, 177
Hiroshima, 177–180
Hisatsune Sakomizu, 177
Homer, 58, 173
Hominids, 1, 2, 4, 5
Homo
    erectus, 5, 9, 11
    habilis, 5, 10
    neanderthaliensis, 2, 3, 6–9
    sapiens, 2, 3, 5, 7–11, 14, 205, 206, 249
    sapiens archaic, 2, 5, 9
Horse, 27, 28, 42, 87, 91, 93, 94, 117, 126
Hot air balloon, 137–139
Hugues de Saint Victor, 112, 113
Huizinga (Johan), 248
Hulot M., 131

Hunaynibn Ishak, 105
Huntington (Samuel P.), 204, 249
Hydraulic
    organ, 50, 54
    pump, 49, 187
Hydrogen balloon, 138, 139
Hydrosphere, 10, 234, 235
Hypatia, 106, 107

I
Ice ages, 9, 14, 21
Imhotep, 32, 33
Imperialism, 152–171, 215
Inclined plane, 23, 24
Industry, 31, 50, 54, 78, 79, 83, 87–90, 125, 133, 136, 137, 184–186, 189, 190, 192–194, 198, 201–203, 220, 224, 228, 230, 234
Information and [Tele]Communications Technologies), 196, 198–200, 228
Innocent II, 127
Innocent IV, 120
Interests on loans, 118, 121
Internet, 180, 198, 222, 229, 230
    of things, 229
Irenaeus of Lyon, 110
Iron age, 6, 17, 34, 69
Irrationalism, 34, 40, 46, 62, 104, 208–210, 223, 249
Islam, 104, 105, 110, 115, 118, 209

J
Jabal, 60
Jaynes (Julian), 40
Jean Buridan, 108, 109
Joachim of Fiore, 113
Job, 27, 30, 31, 54, 62–65, 72, 73, 90, 126, 190
Johannes Scotus Eriugena, 111
John Philoponus, 107, 217
Joliot-Curie (Frédéric), 146
Jubal, 60
Judas Iscariot, 73
Julius Caesar, 174

K
Kaczynski (Theodore), 220, 221
Kant (Immanuel), 168–170
Kardec (Allan), 213, 216
Kennedy (John Fitzgerald), 196
King (Ernest Joseph), 179
Kipling (Rudyard), 203
Koichi Kido, 177
Kurzweil (Ray), 230

L

Lagrange (Joseph-Louis), 151
Lana de Terzi (Francesco), 137
Landing on the Moon, 221
Langley (Samuel Pierpont), 140, 141
Language, 3, 5, 7, 8, 11, 19, 20, 43, 52, 59, 61, 99, 104, 111, 157, 197, 239
Lathe, 130, 131, 201
Leahy (William Daniel), 179
Le Dantec (Felix), 215
Leonardo da Vinci, 122, 129, 137
Leo X, 121
Le Roy (Édouard), 10
Lever, 23–25, 130
Lévi (Eliphas), 213
Life expectation, 73
Light bulb, 144
Lighting gases, 186
Lilienthal (Otto), 140, 142
Lithosphere, 10, 234
Lombe (John), 186
Lomborg (Bjøn), 242, 248
Long (Antony A.), 58
Lope de Vega, 132
Louis the German, 99
Louis XVI, 132
Ludd (Ned), 190
Luddites, 190, 220, 221
Ludwig, 115, 143

M

MacArthur (Douglas), 179
Machine tools, 130–132, 185, 187, 188, 192, 197, 201, 234
Magical thought, 70, 198, 205–208, 213, 221
Malthus (Thomas), 184, 200
Malthusianism, 184, 232
Manchester (William), 97
Marcianus Capella, 111
Marconi (Guglielmo), 144
Marks (Joel), 170
Marquis d'Arlandes, 137, 138
Marshall Plan, 194
Martin (Felix), 36
Marx (Karl), 127, 185, 204
Maxwell (James Clerk), 144, 149
Mechanization, 93, 185, 186, 190, 193
Mechatronics, 194
Meikle (Andrew), 183
Melissus, 44, 45
Memory, 19, 43, 68, 211, 212, 228, 245, 249
Mesmer (Franz Anton), 145
Mesolithic (crisis of), 6, 12–14, 29, 236, 241
Metallurgy, 22, 34, 49, 52, 85–86, 93, 112, 127, 192

Metempsychosis, 67, 211–213
Meteoric iron, 18
Meucci (Antonio), 144
Microelectronics, 228
Middle Ages, 34, 70, 71, 79, 83–85, 87, 93, 97–123, 125–128, 149, 153, 160, 186, 203, 205, 218, 253
Miniaturization, 149, 198, 228–230
Modernization, 204, 249
Moffett (Samuel H.), 104
Monasticism, 110, 113–115, 117, 119
Money, 7, 32, 35–37, 42, 73, 80, 101, 118–120, 155, 159, 160, 162, 167, 194, 195, 205, 239
Monge (Gaspard), 151
Montesquieu (Charles-Louis), 171, 214
Montgolfier
 (Jacques Étienne), 137
 (Joseph Michel), 137
Monti (Vincenzo), 121, 138, 139, 247
Monti di Pietà, 121
Moore (Gordon), 229
Moore's law, 229–231
Morel (Jean-Paul), 80, 84
Morse (Samuel), 144
Moses, 60, 61
Multiculturalism, 204, 223
Mutations, 1, 15, 19, 158, 199, 200, 237
Mystery cults, 207, 211
Myth of the Ages, 66–71, 211

N

Nagasaki, 177–180
Nag Hammadi, 74, 210, 211
Nanotechnologies, 202
Napoleon, 150, 162
Nazism, 170
Neolithic revolution, 10, 14–16, 20, 21, 199, 235–236, 241, 242, 246
Neo-Luddites, 221
Nero, 125
New Age, 69, 208
Newcomen (Thomas), 135, 136
Newton (Isaac), 208
Nietzsche (Friedrich Wilhelm), 215, 216
Nimitz (Chester William), 179
Nirvana, 211
Noble (David F.), 71, 72, 112, 113
Noosphere, 10, 234
Nuclear
 power plant, 146, 148, 233
 reactor, 146, 147, 195
 weapons, 173, 176–180, 195
Nuclear fusion, 149, 225, 227, 233

O

Occultism, 109, 208, 213
Oden (Tinsley J.), 197
Odoacer, 102
Ohm (Georg Simon), 143
Oil, 50, 146, 148, 193, 196, 200, 224–228
Opium wars, 152
Orang-utans, 2
Orestes, 106, 107
Ørsted (Hans Christian), 143
Osborne (Roger), 102
Ouroboros, 67, 210–213

P

Pachomius, 114, 115
Pacioli (Luca), 122
Paleolithic, 5–13, 15, 24, 25, 29, 201, 245
Palladius, 114
Paper, 35, 36, 39, 126, 127, 137, 138, 198, 248, 249
Papyrus, 47, 55
Parabolani, 107
Parchment, 47, 118, 120
Parmenides, 43, 44, 47, 209
Paul III, 168
Paul of Tarsus, 71, 217
Pekáry (Thomas), 54, 78
Pera (Marcello), 223
Pericles, 41, 159
Petrarca (Francesco), 99
Phidias, 32
Philo of Alexandria, 64
Phonograph, 144
Photosynthesis, 227, 235
Pilâtre de Rozier (Jean François), 137, 138
Pius II, 168
Plato, 32, 41, 44, 47, 51, 64–67, 69, 74, 85, 143, 211
Pleroma, 211

Q

Qabbalah, 75
Qoeleth, 63–65
Quantum computers, 231
Quesnay (François), 161, 162
Quevedo (Francisco), 132
Qusta ibn Luqa, 105

R

Radio, 144, 150, 228
Rae (John), 162, 163
Ramelli (Agostino), 91, 129, 130
Rapetti (Anna M.), 115
Ratzinger (Joseph), 223
Reductionism, 133, 197
Renan (Ernest), 215
Renewable energies, 200, 227
Roads, 27, 28, 87–89, 94, 148, 150, 165, 175, 246
Robert
    (Nicolas-Louis), 138
    brothers, 138
Robert Grosseteste, 108
Robespierre (Maximilien), 170, 214
Roger Bacon, 108, 109, 113, 137
Romulus Augustulus, 102
Roosevelt (Franklin Delano), 178
Rostovtzev (Mihail I.), 78, 83, 103
Rousseau (Jean-Jacques), 97, 214
Royal Society of London, 157
Russell (Bertrand), 97
Russo (Lucio), 53, 91
Rutherford (Ernest), 145

S

Sade (Donatien-Alphonse-François), 169
Sadocites, 62, 64, 69
Sadoq, 62
Sartre (Jean Paul), 209
Savery (Thomas), 134–136
Schiavone (Aldo), 78, 79, 81, 83, 103
Schumpeter (Joseph Alois), 163
Science, 7, 32, 40, 41, 44–49, 51–54, 61, 64–66, 68, 70–72, 80, 81, 104, 106–110, 113, 128, 132–134, 136, 139, 143–145, 149, 150, 157–158, 164, 167, 169, 180, 182, 183, 185, 197–199, 203, 207–212, 215, 217–222, 224, 228, 240, 247, 249
Scientism, 215
Scotus Eriugena, 111, 217
Seaborg (Glenn), 146
Second Industrial Revolution, 189, 192–194, 225
Sedentarization, 15
Seeder, 183
Seleucus, 48
Seneca, 52, 80, 83, 85, 90, 111, 143
Shaduf, 24, 25
Shelley (Mary), 145
Siemes (John A.), 178
Singularity, 68, 230, 231
Siphon, 88
Sixtus IV, 168
Slavery, 29, 30, 81, 82, 84, 90, 92, 103, 104, 167, 168, 186
Small (James), 183
Smith (Morton), 64
Socrates, 41, 57, 106
Socrates Scholastic, 106

Sokal (Alain), 198, 221, 222
Solow (Robert Merton), 163
Sophocles, 42
Spaatz (Carl A.), 179
Space economy, 201, 202, 233, 234
Space tourism, 201, 234
Spear thrower, 12, 13
Spencer (Herbert), 204
Spengler (Oswald), 218, 248
Sphericity of Earth, 109
Spiritism, 216
Spiritualism, 213, 214, 217
Stark (Rodney), 59, 70, 71, 78, 104, 105, 168, 203
Steel, 18, 34, 39, 86, 153, 185–187, 192, 203
    industry, 185, 186, 192, 203
Steelworking, 39
Steiner (Rudolph), 211, 212
Steuart (Sir James), 185
Stevenson (Ian), 212
Stilicho, 98, 101
Stone age, 1, 172, 180, 225
Strassmann (Fritz), 145
Stuart Mill (John), 163
Swords, 18, 22, 34, 111, 117, 174
Synesius of Cyrene, 106
Synthetic food, 235, 236
Szilard (Leo), 178

T
Taccola, 129
Tacitus, 82, 86, 165
Taine (Hyppolite), 215
Tamerlane Mongols, 154
Technical drawing, 130, 151, 197
Technosphere, 10, 203, 234
Tehilard de Chardin (Pierre), 10
Telegraph, 144, 149, 174
Telephone, 144
Telescope, 132
Teller (Edward), 178
Templars, 120, 121
Terraforming, 234
Textile industry, 185, 186, 190
Thales, 40, 41, 58, 143
Theatres of machines, 128–130, 137, 181
Theodosius II, 106
Theon of Alexandria, 106
Theosophy, 216, 217
Thermal machines, 92, 136
Third Industrial Revolution, 194–198, 202, 224
Thomas Aquinas, 119, 121, 210
Thomson (William, Lord Kelvin), 136

Thought, 2, 7, 34, 43–45, 57, 58, 61, 64–66, 70, 71, 74, 107, 119, 136–138, 143, 145, 165, 171, 174, 198, 205–213, 220, 221, 230
Thucydides, 46
Timossi (Roberto Giovanni), 169
Torriani (Janello), 131
Torricelli (Evangelista), 134, 137
Townsend (Joseph), 185
Trajan, 84
Transistor, 228, 229
Treadmill, 54, 90, 91, 93
Tubal-Cain, 60
Tucker (Jim B.), 212
Tull (Jethro), 183

U
Ulysses, 30
Upanishad, 207
Uri, 60

V
Valens, 100
Valori (Giancarlo Elia), 214
Valturio (Roberto), 129
Vannuccio Biringuccio, 129
Vehicles, 4, 25–28, 109, 126, 128, 140, 143, 147–149, 181, 191–193, 195, 201, 226, 230, 233, 247
Vernadskij (Vladimir Ivanovič), 10
Verre, 80
Vietnam war, 196, 224
Villard de Honnecourt, 128–130
Virgil, 69, 98, 173
Virtual reality, 199
Vitruvius, 50, 51, 53, 84, 86
Volta (Alessandro), 143
Voltaire, 97
von Guericke (Otto), 134, 137, 143
Vulgars, 53, 99
Vulgata, 59

W
Wachowski (sisters), 212
Wagon, 26, 27, 86, 94, 128, 187
Wallace (Alfred Russel), 185
Water
    mill, 91, 116, 117, 136
    pump, 50, 54, 136
    wheel, 49, 54, 55, 91, 92, 185, 187, 189
Watt (James), 136

Weapons, 2, 12, 13, 17, 18, 22, 34, 55, 74, 86, 127, 146, 171–173, 175–180, 195
   of mass destruction, 171, 173, 175, 176
Weaving, 17, 43, 185, 189
Wedge, 23, 24
Weeber (Karl-Wilhelm), 174
Wells (Peter S.), 100, 103
Wenham (Francis Herbert), 140
Westernization, 204, 224, 249
Wheel, 23, 25–27, 54, 91, 92, 109, 111, 117, 130, 187, 189, 205
   and axle, 23, 25, 26
Whitehead (Gustave), 141
White (Lynn) Jr., 116
William of Occam, 108, 109
William the Conqueror, 117
Windmill, 91, 92, 128, 129
Work, 2, 5, 9, 10, 17, 21, 22, 24, 29–33, 39, 43, 49, 51, 52, 58, 59, 61, 64, 66, 69–74, 80, 83, 84, 99, 100, 103–105, 107, 109, 110, 112–119, 121, 128, 131, 133–135, 137, 140, 151, 156, 158, 160, 169, 174, 181, 184, 185, 193, 203, 207, 208, 217, 219

World population, 242–244
Wright
   (Orville), 142
   (Wilbur), 141
Writing, 19–21, 45, 47, 59, 64, 70, 74, 84, 89, 111, 118, 140, 207, 210, 245, 248

X
Xenophanes, 42

Y
Yahweh, 59, 62, 63, 65, 66
Yamani (Ahmed Zaki), 225, 228
Young (Julian), 220

Z
Zeus, 42, 58
Zheng He, 153, 154, 196
Zilla, 60
Zonca (Vittorio), 129

MIX
Papier aus verantwortungsvollen Quellen
Paper from responsible sources
FSC® C105338

If you have any concerns about our products,
you can contact us on
ProductSafety@springernature.com

In case Publisher is established outside the EU,
the EU authorized representative is:
**Springer Nature Customer Service Center GmbH
Europaplatz 3, 69115 Heidelberg, Germany**

Printed by Libri Plureos GmbH
in Hamburg, Germany